NUTRITIONAL BIOCHEMISTRY

Patricia Trueman

Professor and Head
Department of Biotechnology
Sree Sastha Institute of Engineering and Technology
Chembarampakkam
Chennai

MJP PUBLISHERS
Chennai 600 005

Cataloguing-in-Publication   Data

Trueman, Patricia (1955 –   ).
    Nutritional Biochemistry / by Patricia Trueman. –
Chennai : MJP Publishers, 2007
    xx, 466 p. ; 21 cm.
    Includes Appendix, Glossary, References and Index.
    ISBN 81-8094-031-4 (pbk.)
    1. Biochemistry, Nutrition.
    2. Nutrition I. Title.
    572.4 TRU MJP 028

MJP PUBLISHERS
47, Nallathambi Street
Triplicane
Chennai 600 005

Nutritional Biochemistry

Preface

This book is written to meet the needs of students of Indian Universities pursuing careers in foods, nutrition and allied courses at the undergraduate and postgraduate levels.

The book gives a comprehensive account of nutrients, their functions, metabolism, storage, deficiencies and treatment. An effort has been made to include additional chapters on the interactions between nutrients and the immune system, nutrients and drugs, nutrients and hormones, antioxidants and sports nutrition. The nutrients in foods and the RDA are taken from the manual, *Nutritive Value of Indian Foods,* published by National Institute of Nutrition. At the end of each chapter, questions are provided for the benefit of students preparing for examinations.

The nutrients in the body do not work in compartments individually. There is always an interaction and relationship between the various nutrients. This interaction has been brought out in each chapter.

I hope students using this book find it valuable.

I would appreciate critical comments from the users of this book so that improvements can be made.

Patricia Trueman

ACKNOWLEDGEMENTS

I am grateful to the authors of many publications, journals and books which were used to compile this book.

I am thankful to my students who have helped me in various stages of the development of this book.

I take this opportunity to thank my family members who have supported and encouraged me throughout this project.

I would like to express my sincere thanks to MJP Publishers, without whose support, encouragement and hard work, this publication would not have been possible.

Patricia Trueman

CONTENTS

16. Nutrient–Drug Interaction   379

INTRODUCTION

Nutrition is the science of foods, the nutrients and other substances, their function, and interactions in health and disease. Nutrients are the constituents in food that must be supplied to the body in suitable amounts, and the nutritional state refers to the condition of health of the individual in relation to deficiencies.

FOOD PROVIDES NUTRIENTS

Food, water, and oxygen are life-giving and life-sustaining substances. Food provides both energy and materials needed to build and maintain all body cells. It is important to distinguish between food and nutrients. Food is the source of nutrients. Nutrients are the nourishing substances in food; they are vital for growth, development and the maintenance of body functions.

Nutrition is the study of these nutrients, what they consist of, how the body metabolizes them, and what they finally do in the body to support health.

CLASSES OF NUTRIENTS

Nutrients in food can be organized into six classes.

1. **Carbohydrates** are a mixture of mainly carbon, hydrogen, and oxygen atoms. Carbohydrates are known as sugars and starches. Starches are a storage form of the simple sugar, glucose.

2. **Lipids** (fats and oils) are a mixture of mainly carbon and hydrogen atoms. There are fewer oxygen atoms in lipids than in carbohydrates. The main form of lipid in food is a triglyceride.

3. **Proteins** are a mixture of carbon, hydrogen, nitrogen, oxygen and sometimes sulphur atoms. The building blocks of proteins are amino acids.

4. **Vitamins** are compounds that enable many biochemical reactions to occur, some of which unlock the energy potential of carbohydrates, fats, proteins and alcohol.

5. **Minerals** also play important roles in metabolic reactions. In addition, they form an integral part of the body's structure. Minerals are elements—groups of one or more of the same atom.

6. **Water** nourishes in many ways. It is vital in the body as a solvent and lubricant and as a medium for transporting nutrients and waste.

STATUS OF NUTRITIONAL HEALTH

Theoretically, the body's nutritional status varies for each individual nutrient it needs. The optimal nutritional status for a nutrient is achieved when body tissues have enough of a nutrient for both 1) metabolic functions and 2) surplus stores of it, which can be mobilized during times of increased need.

When nutrient intake does not regularly meet the nutrient needs dictated by cell activities and body maintenance, stores of nutrients soon become depleted, some sooner than the others. Although body stores briefly may be sufficient to compensate for an inadequate diet, over time, serious problems can arise. Components of nutrition assessment are given in Table 1.1.

Biochemical lesions   Once nutrient stores are depleted, a continuing nutrition deficit drains body tissues further. When tissue levels fall far enough, the body's metabolic processes eventually slow down, or even stop. A biochemical lesion then develops. This deficiency is termed subclinical, because there are no outward signs or symptoms.

Clinical lesions   If a biochemical deficit becomes very severe, clinical lesions eventually develop and become outwardly apparent. It is now possible to actually note changes in the body, perhaps in the skin, hair, nails, tongue and eyes. The person also lacks the vigorous glow that good health conveys.

Overnutrition This is overloading the body with too many kilocalories and nutrients. In the long run it results in obesity and its related diseases—diabetes, hypertension and heart disease.

Table 1.1 Components of nutrition assessment

Component	Example
Background history	Medical history including current diseases and past surgeries
	Medication history
	Social history; married/unmarried, cooking facilities, economic status
Nutrition parameters	Anthropometric assessment; height, weight, skin-fold thickness, arm muscle circumference, and other parameters
	Biochemical assessment of blood and urine, enzyme activities
	Clinical assessment, general appearance of skin, eyes and tongue, rapid hair loss, sense of touch, ability to walk
Diet history	Usual intake or record of previous day's meals

FACTORS AFFECTING OUR TASTES

The basic purpose of food is nourishment. It symbolizes much of what we think about ourselves. We bond relationships and friendships around the dinner table.

Early experience Food preferences are learned early as social and cultural preferences. They refine through interactions with parents and friends, our social class, and the need for status. More exposure to a variety of foods at an early age establishes food habits that carry on into old age.

Habit Most of us eat only from a core group of foods. Our food habits are dictated by routine and practice in selecting foods.

Taboos further limit cuisine. Adults even limit the time of the day that foods can be eaten. We do not choose to have soup for breakfast.

Health Several surveys have revealed that about half the population consider nutrition or what they think is good nutrition as an important factor in influencing their food purchases.

Advertising The food industry spends a lot on advertising their product, all of which may not be appropriate.

Social factors Many social changes in recent years have strongly affected the food market place. The increase in the number of working mothers has resulted in foods that can be cooked easily or are pre-cooked.

As education level increases, diet quality also increases but many now tend to skip breakfast.

Economics The amount of money available influences food choices. More affluent people tend to consume more fruits, nuts, fish, meat and poultry.

TERMS FOR UNDERSTANDING NUTRITION

Chemistry A tool for understanding nutrition.

A basic understanding of chemistry can make the study of nutrition much easier. It enables us to connect nutrient characteristics with a mental picture of their structures.

The basic units The body is composed of molecules which are simply clusters of atoms. The term compound refers to a cluster of unlike chemicals grouped into molecules. The basic chemicals of nutrition are carbohydrates, fats, proteins, vitamins, minerals and water.

First molecules form, then many molecules combine to form cells. Cells combine and become tissues which connect to form organs. The interaction of organs forms a coordinated system, called an organism. The human body is just such a coordinated unit of many organ systems.

In its simplest form, an atom has a central core or nucleus, composed of protons and neutrons. Electrons orbit the nucleus. The number of protons equals the number of electrons and together determine the molecular weight. If the number of neutrons differs from the number typically associated with a particular element, the atom exists in an isotope form.

When an atom holds fewer electrons than protons, it exists as a positive ion such as sodium ion (Na^+). When electrons outnumber protons an atom exists as a negative ion such as a chloride ion (Cl^-).

Forming chemical bonds   When two atoms share electrons they create a chemical bond called a covalent bond. Some of the other bonds that exist are the hydrogen bond, ionic bond, etc.

pH   The term pH describes the hydrogen ion (H^+) concentration of a solution. pH is equal to the negative exponent of the hydrogen ion concentration. As the hydrogen ion concentration increases, the pH decreases. Because of the exponential relationships, a one unit change in pH represents a tenfold change in hydrogen ion concentration.

Percentages   The term per cent (%) refers to a part of the total when the total represents 100 parts.

Recommended Dietary Allowances   Recommended dietary allowance (RDA) is defined as the average daily dietary intake level that is sufficient to meet the nutrient requirement of nearly all healthy individuals in a particular life stage and gender group.

Dietary intake, growth studies, nutrient balance, nutrient turnover, and depletion and repletion studies are used to derive RDA.

The RDA serves as a guide to help meet nutritional needs with food rather than with vitamin and mineral supplements.

The RDA is an estimate of the amount of nutrient that should be consumed by healthy people to meet their nutritional need.

Nutrient density   Nutrient density is a concept that has gained acceptance in the last few years.

A way of quantifying nutrient density for a food is to determine its index of nutritional quality (INQ). The INQ is calculated by the following formula.

$$INQ = \frac{\dfrac{\text{Amount of nutrient in 100 g of food}}{\text{RDA for that nutrient}}}{\dfrac{\text{Kilocalories in 100 g of the food}}{\text{RDA for energy}}}$$

An INQ of 1 indicates that the food contributes as much towards meeting the day's need for a specific nutrient as it does for meeting the day's need for kilocalories.

If the INQ is greater than 1, then the food is considered an adequate source for that nutrient. When the INQ is 2 or above, the food is considered a good source of the nutrient. An INQ of 6 indicates the food is an excellent source for that nutrient. An INQ of less than 1 indicates that the food is not a very good source of that nutrient.

No food has INQ values greater than 1 for all nutrients. There is no perfect food. For example, whole milk has a very low INQ value for iron (0.1).

The INQ can be used to find the best foods for supplying any particular nutrient at the least kilocalorie cost.

CARBOHYDRATES

Did You Know?

- Common table sugar is called sucrose. Glucose combined with fructose forms sucrose, or table sugar.

- Carbohydrates are obtained mostly from plants. Few carbohydrates are present in animals foods.

- The primary role of carbohydrates is to supply energy. They are the most important energy-supplying nutrients.

- Fibre and "roughage" are the same thing. Fibre is referred to as bulk, but dietary fibre is the preferred term.

- Milk is a source of carbohydrates. In the exchange system milk yields 12 grams of carbohydrate per exchange.

- Excess dietary carbohydrate is converted into fat in the body. When eaten in excess of energy needs, carbohydrates are stored as glycogen and fat.

- No desirable level of sugar intake has been established but current levels may be too high. The average North American consumes about 125 pounds of simple sugar annually. Many nutritionists believe that this is too high. A diet including less than 50 pounds per year allows for more healthful foods in the diet.

- Honey is not safe to feed to infants. Honey may contain bacterial spores that may grow and become fatal to infants because their stomachs have yet to develop the strong acidic environment that adult stomachs have. Acid reduces bacterium's growth.

- Carbohydrate-loading can help long-distance runners run farther. Researchers have found that if an athlete eats greatly increased amounts of carbohydrates—about 70% of total kcal—each day for 3 days before an event, he can greatly enhance carbohydrate storage in the muscles. This enhances "aerobic" athletic endurance.

Carbohydrates are the most abundant organic chemicals found on earth and synthesized through photosynthesis. They include starch, fibres and sugars.

CLASSIFICATION OF CARBOHYDRATES

Carbohydrate is composed of three elements—carbon, hydrogen and oxygen. The ratio of hydrogen to oxygen in almost all carbohydrates is 2 : 1, the same as in water.

The plant may store the carbohydrate in three major forms: monosaccharides and disaccharides known as simple sugars, and polysaccharides known as complex carbohydrates which include starch, cellulose or fibre and some related compounds.

Monosaccharides

The simplest form of carbohydrate is the monosaccharide. There are three monosaccharides of importance in nutrition namely glucose, fructose and galactose among the hexoses and a few like ribose under the pentoses.

Glucose Glucose is also known as dextrose or blood sugar which is easily found in fruits, vegetables and honey. Glucose is the main source of energy for the central nervous system. The CNS uses about 140 g (about 9 tbsp) of glucose per day. Red blood cells need about 40 g (3 tbsps).

Glucose is the only form in which carbohydrate can be transported in the blood to all the tissues such as muscles, heart and lungs. Hence, it is called as blood sugar. The normal blood glucose level is 80–120 mg/dl. When glucose level of blood increases above 160 mg/dl, the person is described as having hyperglycaemia or too much sugar in blood and when the blood glucose level falls below 60 mg/dl the condition is known as hypoglycaemia. This condition is characterized by weakness, sweating, and light-headedness.

Sorbitol is reduced form of glucose found in apples, pears and peaches and causes flatulence and diarrhoea.

Fructose　It is also known as fruit sugar or levulose. It occurs naturally in many fruits and berries and makes up over one-third of the sugar in honey. Fructose does not cause a rise in blood sugar in the diabetic because insulin is not required for it to be taken up by the cell.

Galactose　This monosaccharide, occurs only as a result of digestion, or hydrolysis of the disaccharide lactose, which is the sugar in milk. Galactose is the only monosaccharide obtained primarily from animal foods. Increased levels of galactose was found to be associated with cataract in animal studies.

Hexoses of physiologic significance is given in Table 2.1.

Pentoses　Ribose is a part of the vitamin riboflavin and the coenzyme derived from it and is also a part of RNA and the DNA (as deoxyribose) (Table 2.2).

Xylose and xylitol are used to provide sweetness and texture in candies without contributing to tooth decay.

Carbohydrates found in glycoproteins are summarized in Table 2.3.

Disaccharides

Sucrose　It is obtained from sugar cane or sugar beet and is used as granulated sugar which is a main sweetening agent in sweets and drinks.

Lactose　It is found exclusively in milk and serves to enhance the absorption of calcium. Lactose has an osmotic effect (a tendency to attract water) which leads to retention of water with a resulting watery stools or diarrhoea.

Some individuals cannot digest lactose due to the absence of the enzyme lactase which is produced in the wall of the intestine. This lactase insufficiency results in lactose intolerance.

Maltose　It is found only in germinating cereals and is a breakdown product of starch.

Table 2.1   Hexoses of physiological importance

Sugar	Source	Importance	Clinical significance
D-Glucose	Fruit juices, hydrolysis of starch, cane sugar, maltose, and lactose	The "sugar" of the body. The sugar carried by the blood and the principal one used by the tissues.	Present in the urine (glycosuria) in the case of Diabetes mellitus owing to increase in blood glucose below (hyperglycaemia).
D-Fructose	Fruit juices, honey, hydrolysis of cane sugar and of inulin (from the Jerusalem artichoke)	Fructose can be converted into glucose in the liver and so can be used in the body.	Hereditary fructose intolerance leads to fructose accumulation and hypoglycaemia.
D-Galactose	Hydrolysis of lactose	Can be converted to glucose in the liver and metabolized. Synthesized in the mammary gland to make the lactose of milk. A constituent of glycolipids and glycoproteins.	Failure to metabolize leads to galactosaemia and cataract.
D-Mannose	Hydrolysis of plant mannans and gums	A constituent of many glycoproteins.	

Table 2.2   Pentoses of physiological importance

Sugar	Source	Biochemical importance	Clinical significance
D-Ribose	Nucleic acids	Structural element of nucleic acids and coenzymes, e.g. ATP, NAD, NADP, flavoproteins. Ribose phosphates are intermediates in pentose phosphate pathway.	
D-Ribulose	Formed in metabolic process	Ribulose phosphate is an intermediate in pentose phosphate pathway.	
D-Arabinose	Gum arabic, plum and cherry gums	Constituent of glycoproteins.	
D-Xylose	Wood gums, proteoglycans, glycosaminoglycans	Constituent of glycoproteins.	
D-Lyxose	Heart muscle	A constituent of a lyxoflavin isolated from human heart muscle.	
L-Xylulose	Intermediate in uronic acid pathway		Found in urine in essential pentosuria.

$$\text{Sucrose} = \text{Glucose} + \text{fructose}$$

$$\text{Lactose} = \text{Glucose} + \text{galactose}$$

$$\text{Maltose} = \text{Glucose} + \text{glucose}$$

Some of the significant disaccharides are listed in Table 2.4.

Table 2.3 Carbohydrates found in glycoproteins

Glycoprotein	Carbohydrates
Hexoses	Mannose (Man), Galactose (Gal)
Acetyl hexosamines	*N*-Acetylglucosamine (GlcNAc) *N*-Acetylgalactosamine (GalNAc)
Pentoses	Arabinose (Ara) Xylose (Xyl)
Methyl pentose	L-Fucose (Fuc)
Sialic acids	*N*-Acyl derivatives of neuraminic acid, e.g. *N*-acetylneuraminic acid (NeuAC), the predominant sialic acid.

Table 2.4 Disaccharides of importance

Sugar	Source	Clinical significance
Maltose	Digestion by amylase or hydrolysis of starch. Germinating cereals and malt.	
Lactose	Milk, may occur in urine during pregnancy.	In lactase deficiency, malabsorption leads to diarrhoea and flatulence.
Sucrose	Cane and beet sugar, sorghum, pineapple, carrot roots	In sucrase deficiency, malabsorption leads to diarrhoea and flatulence.
Trehalose	Fungi and yeasts. The major sugar of insect haemolymph.	

Polysaccharides

These are complex carbohydrates which include starch and dietary fibre.

Starch   Starch is a polysaccharide made of glucose units either as one long chain (amylose) or in a branched arrangement (amylopectin) found in cereals, potatoes and other roots and tubers.

Glycogen   The animal stores carbohydrates as glycogen primarily in the liver and muscle. The energy stored in glycogen represents only enough energy to last an adult human being for about half a day.

Fibre   There are two types of fibre found in food namely crude fibre and dietary fibre. Crude fibre is that portion of a plant that resists chemical breakdown, and dietary fibre is any material that remains undigested in the intestine.

The starches are also classified based on their reaction and rate of digestion within the digestive tract (Table 2.5).

Table 2.5   Classification of starch for nutritional purposes

Type of starch	Occurrence	Digestion in small intestine
Rapidly digestible starch (RDS)	Freshly cooked starchy food	Rapid
Slowly digestible starch (SDS)	Most raw cereals	Slow but complete
Resistant starch (RS)		
Physically inaccessible starch	Partly milled grains and seeds	Resistant
Resistant starch granules	Raw potato and banana	Resistant
Retrograded amylose	Cooked potato after cooling, bread, cornflakes, etc.	Resistant

Rapidly digestible starch (RDS) consists mainly of amorphous and dispersed starch. It is typically found in large amounts in starchy foods that have been cooked by moist heat.

Slowly digestible starch (SDS), like RDS, is expected to be completely digested in the small intestine, but digested very slowly. This category includes starch that is poorly accessible to enzymes.

Resistant starch (RS) comprises the starch that may potentially resist digestion in the small intestine. These are highly resistant to hydrolysis.

The amounts of RDS, SDS and RS in foods are highly variable. They depend partly on the source of starch but largely on the type and extent of processing the food has undergone.

Dextrins are degradation products of starch in which the glucose chains have been broken down to smaller units by partial hydrolysis. Liquid glucose is a mixture of dextrins, maltose, glucose and water (Table 2.5).

FUNCTIONS

1. **Source of energy**   The primary function of carbohydrates in human nutrition is to provide fuel for energy. Carbohydrates help to maintain energy reserve.

   The liver and muscle glycogen reserves provide a constant interchange with the body's overall energy balance system and protect cells from depressed metabolic function and injury.

   Glycogen is stored in liver and muscle. Stored amounts are given in Table 2.6.

Table 2.6 Glycogen and sugar content of normal man (Body wt. = 70 kg, Liver wt. = 1800 g, Muscle mass = 35 kg, Volume of blood and ECF = 21 l)

	g/kg	Total g
Muscle glycogen	7	245
Liver glycogen	60	108
Blood and ECF sugar	0.8	17
Total body carbohydrate		370

2. **Protein sparing action** Carbohydrate helps to regulate protein metabolism. The presence of sufficient carbohydrates for energy demands of the body prevents the channelling of too much protein for this purpose. This protein sparing action of carbohydrate allows a major portion of protein to be used for its basic structural purpose of tissue building.

3. **Antiketogenic effect** Carbohydrates also relate to fat metabolism. The amount of carbohydrate in the diet determines how much of fat will be broken down thus affecting the formation and disposal of ketone bodies. When too much of fats are oxidized (in the absence of carbohydrates) it results in the accumulation of ketone bodies which lead to a condition called ketoacidosis.

4. **Role in the heart** Glycogen in heart muscle is an important emergency source of contractile energy. In a damaged heart, poor glycogen stores or low carbohydrate intake may cause cardiac symptoms and angina.

5. **Role in the central nervous system** The brain contains no stored supply of glucose and is therefore especially dependent on a minute-to-minute supply of glucose from the blood.

6. **Role in muscle** Carbohydrates are the major source of energy for muscular work.

7. **Role in liver**   It is involved in detoxifying action.

8. **Synthesis of nucleic acids**   Ribose (a pentose sugar) is necessary for the synthesis of RNA and DNA.

DIGESTION

Digestion of polysaccharides begins in the mouth. Table 2.7 provides a summary of digestion of carbohydrates.

Table 2.7   Summary of carbohydrate digestion

Organ	Enzyme	Action
Mouth	Ptyalin	Starch → dextrins → maltose
Stomach	None	Above action continues till medium becomes acidic
Small intestine	Pancreatic amylase	Starch → dextrins → maltose
	Intestinal	
	Sucrase	Sucrose → glucose + fructose
	Lactase	Lactose → glucose + galactose
	Maltase	Maltose → 2 glucose

ABSORPTION

Carbohydrates are absorbed into the bloodstream as glucose, galactose and fructose. The simple sugars enter the portal circulation and are transported to the liver. Here the fructose and galactose are converted to glycogen for storage.

The absorption of glucose is affected by the amount of sodium ions in the intestinal lumen. A high Na^+ concentration facilitates glucose influx into the intestinal mucosal cell whereas a low Na^+ concentration inhibits influx. This is because glucose and Na^+ share the same co-transporter carrier protein. Glucose moves along with the Na^+ and is released inside the cell. The glucose transport mechanism also transports galactose. Fructose is transported via a different carrier and its absorption is independent of Na^+. The rate of absorption of glucose and galactose is almost equal but fructose is absorbed half as rapidly.

Factors Affecting Absorption

Absorption is influenced by 4 factors.

- The rate at which the carbohydrates enter the small intestine affects its absorption. The rate of entry depends upon the mobility of the stomach and the control of the duodenal sphincter muscle, the pyloric valve.

- The type of the food mixture present influences the degree of competition for absorptive sites and the available carrier transport system.

- Absorption is also influenced by the condition of intestinal membrane and the time for which the carbohydrate is held in contact with the membrane. Any abnormal mucosal tissue or an abnormally rapid movement of the carbohydrate along the intestine (diarrhoea) hinders absorption.

- Normal endocrine activity of the anterior pituitary and the related functioning of the thyroid is necessary for normal absorption. In addition, the adrenal cortex hormones regulate the body sodium exchange, which indirectly influences the operation of the sodium pump.

CARBOHYDRATE METABOLISM

Before pyruvate can enter the citric acid cycle, it must be transported into the mitochondrion via a special pyruvate transporter that aids its passage across the inner mitochondrial membrane. This involves a symport mechanism whereby one proton is co-transported. Within the mitochondrion, pyruvate is oxidatively decarboxylated to acetyl-CoA. This reaction is catalysed by several different enzymes working sequentially in a multienzyme complex. They are collectively designated as the pyruvate dehydrogenase complex (Figure 2.1). Pyruvate is decarboxylated in the presence of thiamine diphosphate to a hydroxy ethyl derivative of the thiazole ring of enzyme-bound thiamine diphosphate, which in turn reacts with oxidized lipoamide to form acetyl lipoamide and then finally acetyl-CoA. The reaction is summarized in Figure 2.2. Table 2.8 shows the energy produced in the glycolytic pathway.

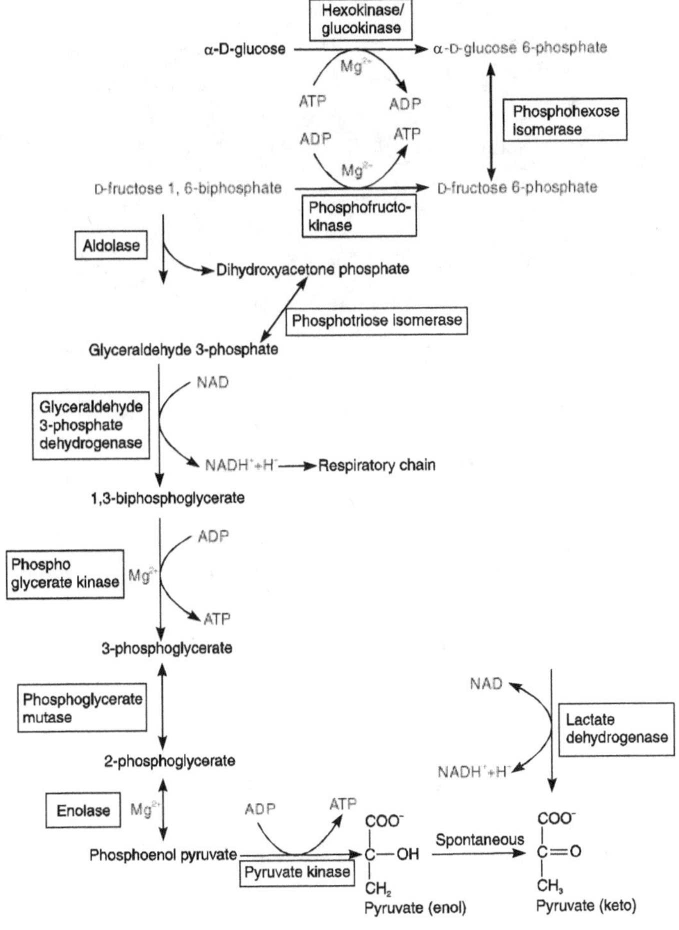

Figure 2.1   Glycolytic pathway

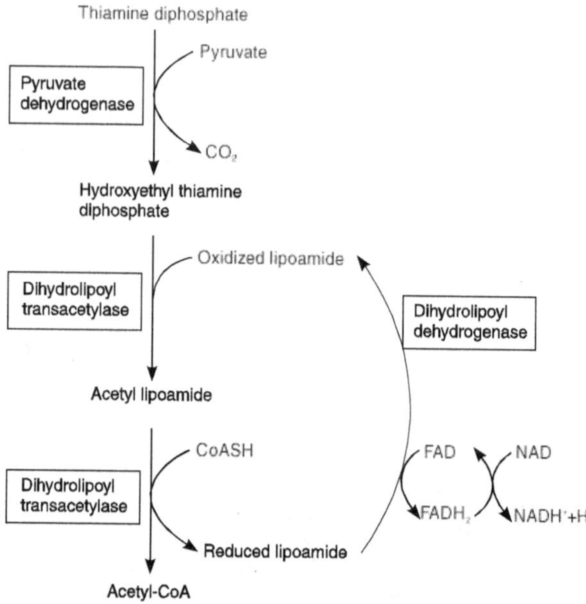

Figure 2.2   Conversion of pyruvate to acetyl-CoA

Table 2.8   Energy produced in the glycolytic pathway

Pathway	Reaction catalysed by	Method of ~P production	No of ~P produced
Glycolysis	Glyceraldehyde-3-PO₄ dehydrogenase	Respiratory chain oxidation of 2 NADH⁺	6
	Phosphoglycerate kinase	Oxidation at substrate level	2
	Pyruvate kinase	Oxidation at substrate level	2
Consumption of ATP by hexokinase and phosphofructokinase			−2
		Net	8

The citric acid cycle (Krebs cycle, TCA cycle) is a series of reactions in mitochondria that bring about the catabolism of acetyl residues. Liberating hydrogen equivalents, which upon oxidation lead to the release of most of the free energy tissue fuels (Figure 2.3).

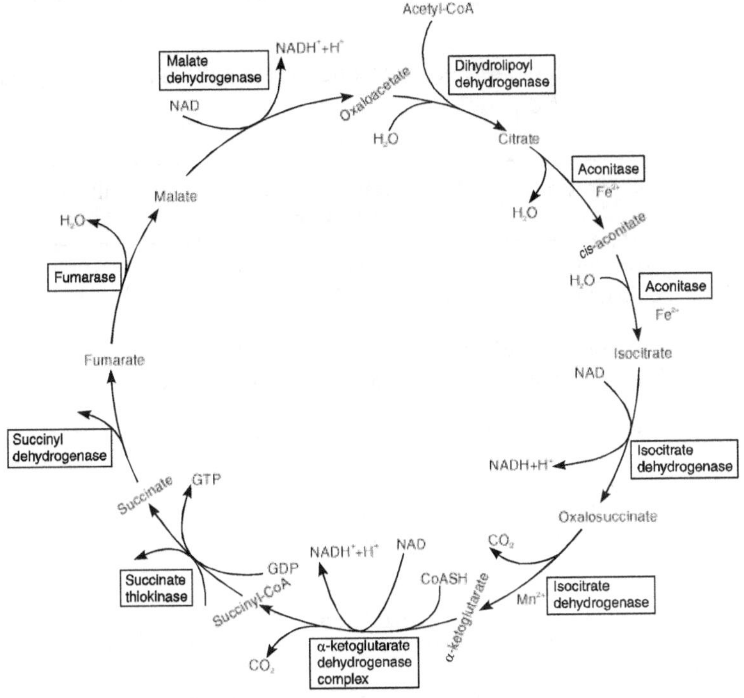

Figure 2.3   Critic acid cycle

The major function of the citric acid cycle is to act as the final common pathway for the oxidation of carbohydrates, lipids and proteins, since glucose, fatty acids and many amino acids are all metabolized to acetyl-CoA or intermediates of the cycle. It also plays a major role in gluconeogenesis, transamination, deamination and lipogenesis. While several of these processes are carried out in most tissues, the liver is the only tissue in which all occur.

The enzymes of the citric acid cycle are located in the mitochondrial matrix, either free or attached to the inner surface of the inner mitochondrial membrane, which facilitates the transfer of reducing equivalents to the adjacent enzymes of the respiratory chain, situated in the inner mitochondrial membrane.

The total amount of energy produced is summarized in Table 2.9.

Table 2.9 Generation of high-energy phosphate bonds by the citric acid cycle

Reaction catalysed by	Method of ~P production	Number of ~P formed
Isocitrate dehydrogenase	Respiratory chain oxidation of NADH	3
α-ketoglutarate dehydrogenase	Respiratory chain oxidation of NADH	3
Succinate thiokinase	Oxidation at substrate level	1
Succinate dehydrogenase	Respiratory chain oxidation of $FADH_2$	2
Malate dehydrogenase	Respiratory chain oxidation of NADH	3
	Net	12

Role of Vitamins in the Citric Acid Cycle

Four of the soluble vitamins of the B-complex have precise roles in the functioning of the citric acid cycle. They are

1. Riboflavin in the form of flavin adenine dinucleotide (FAD) a cofactor in the α-ketoglutarate dehydrogenase complex and in succinate dehydrogenase.

2. Niacin in the form of nicotinamide adenine dinucleotide (NAD), the coenzyme for three dehydrogenases in the cycle, isocitrate dehydrogenase, α-ketoglutarate dehydrogenase, and malate dehydrogenase.

3. Thiamine (B_1) as thiamine diphosphate, the coenzyme for decarboxylation in the α-ketoglutarate dehydrogenase reaction.

4. Pantothenic acid as part of coenzyme A, the cofactor attached to "active" acyl residues such as acetyl-CoA and succinyl-CoA.

When 1 molecule of glucose is combusted in a calorimeter to CO_2 and water, approximately 2780 kJ are liberated as heat. When oxidation occurs in the tissue, some of this energy is captured as high-energy phosphate bonds. Assuming each high-energy bond to be equivalent to 30.5 kJ the total energy captured in ATP per mole of glucose is 1159 kJ or approximately 41.7% of the energy of combustion.

Anabolic Pathways

Gluconeogenesis includes all mechanisms and pathways responsible for converting non-carbohydrates to glucose or glycogen. The major substrates for gluconeogenesis are the glucogenic amino acids, lactate, glycerol, and in ruminants, propionate. Liver and kidney are the major tissues involved, since they contain a full complement of the necessary enzymes.

Gluconeogenesis meets the needs of the body for glucose when carbohydrate is not available in sufficient amounts from the diet. A continual supply of glucose is necessary as a source of energy, especially for the nervous system and the erythrocytes. Below a critical blood glucose concentration, there is brain dysfunction, which under conditions of severe hypoglycaemia can lead to coma and death.

Gluconeogenesis is a modification and adaptation of the glycolysis pathway and the citric acid cycle (Figure 2.4).

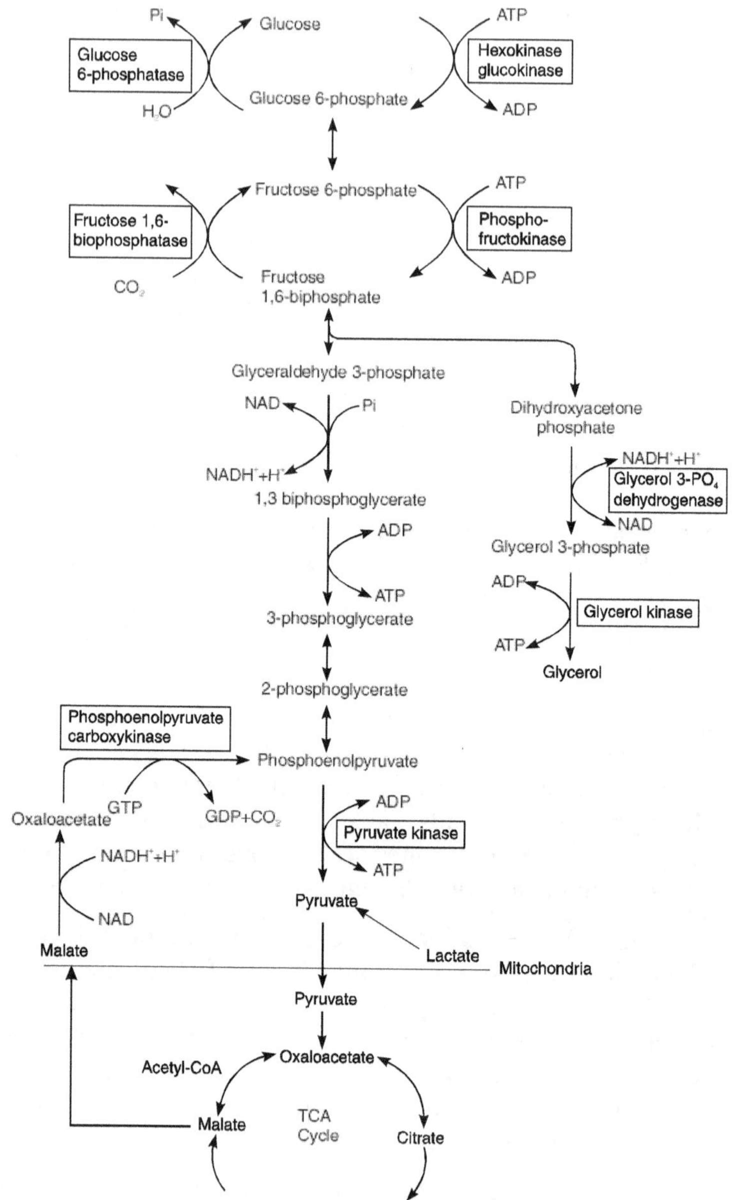

Figure 2.4   Gluconeogenesis

Energy barriers obstruct a simple reversal of glycolysis 1) between pyruvate and phosphoenolpyruvate 2) between fructose 1,6-biphosphate and fructose 6-phosphate 3) between glucose 6-phosphate and glucose and 4) between glucose 1-phosphate and glycogen. These barriers are circumvented by special reactions.

Present in mitochondria is an enzyme, pyruvate carboxylase, which in the presence of ATP, the B vitamin biotin and CO_2 converts pyruvate to oxaloacetate. The function of biotin is to bind CO_2.

Oxaloacetate does not diffuse readily from mitochondria. However malate diffuses through the membrane easily and reforms oxaloacetate in the extramitochondrial part in the presence of malate dehydrogenase. Oxaloacetate in turn is converted to phosphoenolpyruvate in the presence of phosphoenolpyruvate carboxykinase. High-energy phosphate in the form of GTP is required in this reaction and CO_2 is liberated.

DIETARY FIBRE

Dietary fibre is defined as that portion of food derived from plant cells, which is resistant to hydrolysis/digestion by the elementary enzyme system in human beings. It consists of hemicellulose, cellulose, lignins, oligosaccharides, pectins, gums and waxes.

Some bacteria in the large intestine can degrade some components of fibre releasing products, that can be absorbed into the body and used as energy source.

Two categories of fibre are found in food. Crude fibre is defined as the residue remaining after the treatment with hot sulphuric acid, alkali and alcohol. The main component of crude fibre is cellulose. Crude fibre is a component of dietary fibre. Pectins, hemicellulose, lignins, gums and mucilages are also found in foods and are also resistant to digestion. These together with cellulose are collectively known as dietary fibre.

In 1969 Burkitt explained the importance of fibre.

Chemistry of Fibre

1. Cellulose provides the chief constituent of the framework of plants. Human beings cannot digest cellulose because they lack the necessary digestive enzymes. Therefore it remains in the digestive tract and contributes important bulk to the diet. This bulk helps move the digested food mass along and stimulates peristalsis. Cellulose makes up the principal structural material in plant cell walls and provides most of the substance labelled "crude fibre". The main sources are stems and leaves of vegetables, seed and grain coverings, skins and hulls.

2. Non-cellulose polysaccharides are carbohydrates that include hemicellulose, pectins, gums and mucilages and algal substances. They absorb water, slow gastric emptying time, bind bile acids and prevent intraluminal colon pressure by providing bulk for normal intestinal muscle action.

3. Lignin is the only non-carbohydrate type of fibre with a large compound forming the woody part of plants. It combines with bile acids to form insoluble compounds in the intestines, thus preventing their absorption.

Dietary fibre produces various effects on the food mix consumed and its fate in the body. Most of these effects are caused by its physiological properties which include the following.

1. *Water absorption* which contributes to its bulk-forming laxative effect and influences the transit time of the food mass through the digestive tract and consequent absorption of the various nutrients in the food mix.

2. *Binding effect characteristic* of certain fibres such as the non-cellulose materials which influence blood lipid levels through their capacity to bind bile salts and cholesterol and prevent their absorption.

3. *Colon bacteria relation* which provides fermentation substrates for bacterial action, providing volatile fatty acids and gas.

Function of Fibre

1. *Physiological value*   (a) Dietary fibre holds water, so stools are soft, bulky and readily eliminated. High fibre intake prevents or relieves constipation and thus acts as a "natural laxative." (b) They also cause dilution of colon contents and potential toxic substances are eliminated.

2. *Retardation of gastric emptying*   This has a beneficial effect in that there is an increased satiety. So less food is eaten, thus helping to keep energy intake low and also adds bulk to the food.

3. *Retardation of intestinal transit time*   The fibres are not hydrolysed and so they stay for a longer time in the intestines and form the bulk of the stools. This also prevents the direct exposure of any toxic substance to mucosa.

    High fibre diets have low coefficients of digestibility and thus the net energy absorbed by the body is less than that from diets of animal foods.

4. *Binding of toxic substances*   The indigestible polysaccharide present in the diet dilutes the toxic substances and acts as a detoxifying agent which is in turn eliminated.

5. *Binding of cholesterol*   The pectins present in the food combine with the cholesterol and thus prevent it from being absorbed. They bind bile acids and cholesterol and help in lowering serum cholesterol.

6. *Effect on diabetes mellitus*   When a high fibre diet is taken there is a feeling of satiety even when low amounts of food is consumed. This is due to decreased gastric emptying and there is a decrease in glucose absorption and the level of blood sugar does not increase too rapidly.

7. *Binding of carcinogens*   Certain carcinogens present in foodstuff are hydrolysed and combine with fibres and are removed from the colon and prevents colon cancer.

8. *Reduction of intraluminal pressure* It reduces intraluminal pressure as the stools are easily eliminated and thereby prevents haemorrhoids (piles).

9. *Decrease in diverticulitis incidence* Dietary fibre leads to the decrease in the incidence of diverticulitis which is weakening of the walls of the intestine due to the pressure of the hard stools. Diverticulitis is the inflammation of the intestine.

10. *Delay of hunger pangs* Dietary fibre delays hunger pangs and can therefore be used to control obesity.

Advantages of fibre consumption in controlling disease is given in Table 2.10.

Disadvantages

1. It causes abdominal discomfort, and gas production in colon due to bacterial action.

2. Increases the excretion of minerals like calcium, iron, etc. which leads to various deficiency diseases.

Table 2.10   Advantages of fibre consumption in preventing diseases

Diseases	Type of fibre involved	Physiological mechanisms
Constipation	Insoluble fibre	Increases water holding capacity
Diverticulitis	Cellulose	Increases stool weight
Irritable bowel syndrome	Bran	Reduces transit time
		Faster bowel emptying due to increased intraluminal mass.
Piles		Decreases intracolonic pressure
Coronary heart disease	Soluble fibre, β-glucan content, oats, skin, pectin	Cholesterol synthesis is inhibited by acetic, propionic and butyric acids produced by bacterial fermentation.
		Clearance of LDL cholesterol

(Contd.)

Table 2.10   (Continued)

Diseases	Type of fibre involved	Physiological mechanisms
Gall stones	Guar gum, fruits and vegetables	Increases excretion of steroids
		Reduces fatty acid absorption
		Fibre binds faecal bile acids and increases excretion of bile-acid-derived cholesterol.
Diabetes mellitus	Soluble fibres, legume seed coverings	Rate of glucose absorption is decreased
Obesity	Soluble fibres	Gastric emptying is delayed and feeling of satiety is increased.
		Diets high in fibre are low in calories.

Dietary fibre content of some common Indian foods is given in Table 2.11.

Table 2.11   Dietary fibre content of some common Indian foods

Food	Dietary fibre (g/100g)
Cereals and Millets	
Rice	7.6
Wheat	17.6
Sorghum	14.3
Bajra	20.3
Ragi	18.6
Pulses and legumes	
Green gram	13.5
Black gram	14.3
Red gram	14.1
Bengal gram	13.6

(Contd.)

Table 2.11   (Continued)

Food	Dietary fibre (g/100g)
Nuts and oilseeds	
Almond nut	6.1
Coconut dry (copra)	8.9
Roots and tubers	
Sweet potato	7.3
Potato	4.0
Yam	5.3
Fruits	
Banana	2.5
Mango	2.3
Vegetables	
Amaranth	3.4
Palak	5.0
Brinjal	2.0
Ridged gourd	5.7
Snake gourd	1.8
Bottle gourd	2.8
Yellow pumpkin	0.5

Source   B S. Narasinga Rao, *Nutrition Foundation of India Bulletin*, 9:(4), 1988.

MAINTENANCE OF BLOOD GLUCOSE LEVEL

Blood sugar refers to the glucose present in blood. It is the only sugar present at all times in the blood, biological fluids and tissues in physiologically significant amounts, though other monosaccharides are also absorbed in the same manner as glucose. The concentration of glucose in blood is remarkably constant in a normal healthy individual, ranging from 80–120 mg% (normoglycaemia). This value may be slightly low during post-absorptive state ranging from 75 to 90 mg%, but it does not reach values less than 70 mg% even during starvation. The blood glucose level during post-absorptive period is referred to as fasting blood sugar. Similarly during the period of maximum absorption after a carbohydrate meal, the blood glucose may increase slightly from 130 to 180 mg% but not beyond 180 mg%. This shows that in a normal healthy individual the blood glucose is maintained at a fairly constant level, in spite of factors operating to alter it. As there is a constant intake and removal of glucose from the body, the blood glucose level at a given time is the resultant of two factors mainly,

1. The amount of glucose entering the bloodstream

2. The amount of glucose removed from the bloodstream

FACTORS AFFECTING BLOOD SUGAR LEVEL

In the body, there are mechanisms operating to maintain the concentration of blood glucose at a fairly constant level; they are considered under the following headings:

1. Alimentary mechanism

2. Role of liver

3. Role of hormones

4. Role of muscles and tissues

5. Role of kidneys

6. Role of blood glucose

7. Formation of fat

Alimentary Mechanism

Following administration of glucose up to 200 g orally, no glucose is detected in the urine and this is due to complete reabsorption of glucose by renal tubules. Increasing the amount of glucose up to 300 to 500 g, will cause distension of the stomach and consequently slow emptying and slow absorption from the intestines. In this way, rapid absorption causing a high rise of blood glucose is prevented. Though the glucose level may increase to some extent, it does not go beyond the renal threshold value. The maximum limit of absorption of glucose is 1.84 g/kg body weight per hour. Thyroxine and adrenocortical hormones increase absorption of glucose, whereas deficiency of vitamin B delays absorption. Other monosaccharides such as galactose and fructose, absorbed along with glucose from the diet are rapidly converted to glucose.

Role of Liver

Liver plays an important role in the regulation of blood glucose level. When the blood glucose tends to rise, liver helps in restoring the normal value by

i. accelerating glycogenesis

ii. increasing utilization of glucose

The excess glucose is converted to glycogen and stored in the liver; or if there is a need for energy, glucose is oxidized and utilized.

Insulin favours both glycogenesis and utilization of glucose (Figure 2.5).

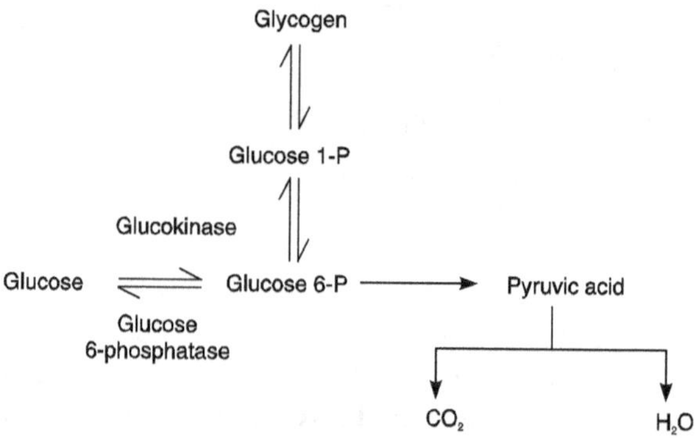

Figure 2.5   Utilization of glycogen and glucose

When the blood glucose tends to fall, in fasting or during intake of low carbohydrate diet, liver helps in restoring normal value by

i. accelerating glycogenolysis

ii. accelerating gluconeogenesis

iii. depressing utilization of glucose

Glycogenolysis is the conversion of liver glycogen to blood glucose. This causes rise in blood glucose. Adrenaline, glucagon and thyroxine favour hepatic glycogenolysis.

Gluconeogenesis is the conversion of non-carbohydrate substances to glycogen and glucose. The non-carbohydrate sources include glucogenic amino acids and certain intermediates of carbohydrate metabolism namely lactic acid, pyruvic acid and glycerol of fat. They enter the pathway of carbohydrate metabolism at various stages and get converted to glucose.

Role of Hormones

Hormones play a very important part in regulating the level of blood glucose to normal limits. The blood glucose of an individual is the result of the opposing action of two sets of hormones namely insulin acting on one hand and hormones of pituitary, adrenal cortex, medulla, and thyroid acting on the other. Variations in blood glucose value occurs leading to abnormal carbohydrate metabolism, if there are disturbances in the secretion of these hormones. A brief account of the action of each of these hormones in given below:

Insulin This hormone is secreted by the β-cells of Langerhans, and is the strongest blood sugar lowering factor. It lowers blood sugar level by

- accelerating glycogenesis in the liver

- increasing the rate of oxidation of glucose by tissue cells for purpose of energy production

ⓢ depressing gluconeogenesis, thus reducing the rate of addition of glucose to the blood from other non-carbohydrate substances

Diabetes mellitus which is characterized by abnormally high blood glucose levels (hyperglycaemia) is a chronic disorder caused by decreased secretion of insulin, due to degeneration of β-cells of Langerhans.

Mechanism of insulin action

i. Insulin promotes hexokinase action which is the initial reaction of glucose oxidation whereby glucose is converted to glucose 6-phosphate. Adrenocortical hormones and hormones of pituitary inhibit this reaction, while insulin is antagonistic by promoting both oxidation and glycogenesis.

ii. Insulin promotes the oxidation of sugars in Krebs citric acid cycle, the oxidation of sugars at the site of acetyl-CoA formation.

iii. Insulin depresses the action of glucose 6-phosphatase of liver so that less glucose is formed by hepatic glycogenolysis.

iv. Insulin increases cell membrane permeability and thus facilitates entry of glucose into cells for oxidation.

Glucagon This is a hormone secreted by the α-cells of Langerhans. Hypoglycaemia stimulates its secretion which in turn promotes hepatic glycogenolysis, raising the blood sugar level. Glucagon is also known as hyperglycaemic glycogenolytic factor (HGF).

Hormones of the adrenal cortex 11-oxygenated adrenocortical hormones are diabetogenic in action. They increase the blood sugar level by, increasing gluconeogenesis from tissue proteins, and diminishing the utilization of glucose by enhancing the inhibitory influence of growth hormone on the hexokinase reaction. These reactions are antagonistic to that of insulin, but similar to that of anterior pituitary hormones.

Hormones of adrenal medulla   Adrenaline (epinephrine) is diabetogenic in action. It increases the blood sugar level by increasing glycogenolysis and glycolysis. Thus it accelerates breakdown of both hepatic and muscle glycogen.

Hormones of the pituitary   These include growth hormone, adrenocorticotropic hormone (ACTH) and TSH.

   i.   Growth hormone is the strongest blood sugar raising factor and therefore diabetogenic in action. Its action is direct, not mediated by gland, and opposite to that of insulin. It acts as follows:

- It raises blood sugar level.

- It decreases utilization of glucose by tissues. This is due to inhibition of hexokinase reaction.

- It increases gluconeogenesis.

- It depresses hepatic and muscle glycogenesis.

   ii.   ACTH and TSH   stimulate the respective organs namely adrenal cortex and thyroid, to secrete adrenal cortical hormone and thyroxine which produce diabetogenic effects. They decrease utilization of glucose and cause hyperglycaemia.

Hormone of thyroid   Thyroxine is also diabetogenic. It increases blood sugar by

   i. Increasing the intestinal absorption of glucose,

   ii. Increasing hepatic glycogenolysis,

   iii. Increasing utilization of glucose.

Role of Muscle and Tissue

Muscle and tissues behave like liver. Glucose is drawn from blood for muscular contraction and tissue activity. When blood glucose is high, muscle and tissues withdraw the glucose and store it as glycogen or convert it into depot fat, thus reducing the blood glucose. When the blood glucose is low, after severe muscular exercise, lactic acid which is formed is withdrawn from the tissues and muscles by the liver and is converted to glycogen and finally enters the blood as glucose (*see* Cori's cycle) (Figure 2.6).

Role of Kidney

It is the last outpost in blood glucose regulating mechanism. Kidneys excrete blood glucose above the renal threshold value (about 180 mg). Glucose is usually reabsorbed by the renal tubules.

Role of Blood Glucose

The feedback mechanism either causes the release of insulin or glucagon.

Formation of Fat

Excess glucose is disposed off by formation of fat. Synthesis of fat from carbohydrate takes place readily in adipose tissues and mammary gland, and for this insulin is required.

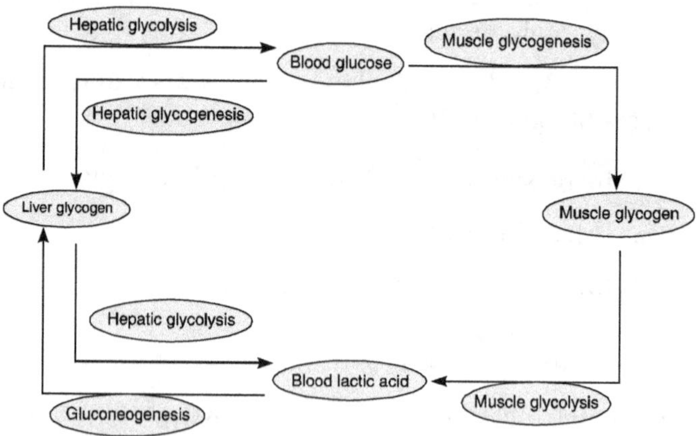

Figure 2.6   Cori's cycle

○ REVIEW QUESTIONS

1. Explain the functions of carbohydrates.

2. What is dietary fibre?

3. Explain the role of fibre in human nutrition.

4. How is blood sugar level maintained?

5. What are non-starch polysaccharides?

6. What is glycaemic index of foods?

7. How are carbohydrates digested and absorbed?

CRITICAL THINKING QUESTIONS ○

1. Are carbohydrates necessarily fattening?

2. Is there an RDA for carbohydrate?

3. What does the human body use dietary fibre for?

 3

LIPIDS

Did You Know?

- A lipid has more kilocalories per gram than alcohol. Alcohol yields 7 kilocalories per gram and lipid yields 9 kilocalories per gram.

- A high serum cholesterol level is a risk factor for stroke.

- Cholesterol is found only in animal-derived foods. But don't be misled by advertisers who state that certain products are "cholesterol-free." The product may still contain saturated fat, which is a more important determinant of blood cholesterol level.

- Animal fat is the major dietary factor that raises serum cholesterol levels.

- Triglycerides are the main form of fat found in both foods and the body.

- Hydrogenation makes vegetable oils more solid at room temperature. In some cases this process makes food production easier; hydrogenation prevents the oils in peanut butter from separating during storage.

- Fruits are essentially fat-free. Most fruits contain only a trace amount of fat.

- Antioxidants protect foods from turning rancid. BHA and BHT are examples of antioxidants found in salad dressings.

- The small intestine absorbs some vitamins better when dietary fat is present. Because vitamins A, D, E, and K are fat-soluble, their absorption is enhanced by dietary fat.

- Butter and margarine contain about the same amount of fat. Butter, however, contains more saturated fats than margarine.

- High doses of fish oils in the diet lower serum cholesterol levels. Studies with both animals and humans have shown that high doses of fish oil can moderately lower serum cholesterol levels. However, high fish oil doses affect serum triglyceride levels.

Fat, like carbohydrate, is composed of three elements carbon, hydrogen and oxygen. However, in fat, the ratio of oxygen to carbon and hydrogen is much lower than in simple carbohydrates. The lower amount of oxygen in relation to the other two elements results in fat being a more concentrated source of energy than carbohydrate.

CLASSIFICATION OF LIPIDS

Simple Lipids

Simple lipids include oil and fats, and waxes.

1. **Oils** and **fats** are esters of fatty acids and glycerol. Oils are liquids at 20°C while fats are solids at 20°C.

2. Waxes are esters of fatty acids and long-chain aliphatic alcohols.

Compound Lipids

Compound lipids contain, in addition to fatty acids and glycerol, some other organic compounds.

1. **Phospholipids** These contain phosphoric acid and nitrogenous base in addition to glycerol, and fatty acids (lecithin, cephalins, plasmalogen).

2. **Sphingolipids** These contain the base sphingosine or dihydrosphingosine.

3. **Glycolipids** These are complex lipids containing carbohydrates in combination with fatty acids and sphingosine.

4. **Sulpholipids** These contain sulphuric acid in combination with hexose in a cerebroside.

5. **Derived lipids** These include fatty acids, alcohols, sterols, carotenoids, etc.

FORMATION OF TRIGLYCERIDES

Dietary fat is composed of two major components: a three-carbon molecule known as glycerol, with one to three fatty acids attached to it. The most common fats both found in food and stored in the body are triglycerides.

$$H_2C - OOC(CH_2)_{14}CH_3$$
$$|$$
$$HC - OOC(CH_2)_{14}CH_3$$
$$|$$
$$H_2C - OOC(CH_2)_{14}CH_3$$

Triglyceride

Saturation

In fatty acid chains, each carbon atom between the methyl end and the carboxyl end has the capability of holding or having two hydrogens attached to it the maximum number possible.

Some fats found in foods do not have the maximum number of hydrogen (have a double bond). These are unsaturated fatty acids. Fatty acids with one double bond in the structure are monounsaturated fatty acids. When there are two or more double bonds, the fatty acid is said to be polyunsaturated fatty acids (PUFA).

There have also been identified very long-chain (22 to 24 carbon atoms) fatty acids with five or six double bonds, and called highly unsaturated fatty acids (HUFA) (Table 3.1).

The degree of saturation is often expressed as the p/s ratio—the ratio of polyunsaturated fatty acids to saturated fatty acids after monounsaturated fatty acids have been excluded.

Monounsaturated fatty acids also have beneficial health effects. Hence the ratio is P-M/s ratio. Where the ratio of polyunsaturated fatty acids plus monounsaturated fatty acids to saturated fatty acids is used to express the health value of a fat.

Dietary triglycerides contain a mixture of fatty acids with varying degrees of saturation and varying chain lengths. Because

Table 3.1   Fatty acids and their sources

Name of fatty acid	No. of carbon atoms	No. of double bonds	Food sources
Saturated			
Short-chain			
Butyric	4	–	Butter
Caproic	6	–	Butter
Caprylic	8	–	Coconut oil
Medium-chain			
Capric	10	–	Palm oil
Lauric	12	–	Coconut oil
Myristic	14	–	Butterfat, nutmeg, coconut oil
Long-chain			
Palmitic	16	–	Animal fat and vegetable oil
Stearic	18	–	Animal fat and vegetable oil
Arachidic	20	–	Peanut oil and lard
Unsaturated			
Long-chain			
Palmitoleic	16	1	Butter and seed oils
Oleic	18	1	Most fats and oils
Linoleic	18	2	Seed fats, corn, etc.
Linolenic	18	3	Soybean oil
Arachidonic	20	4	Peanut oil, lard
Eicosapentaenoic acid (EPA)	20	5	Fish oil
Docosahexaenoic acid (DHA)	20	6	Fish oil

glycerol is common to all triglycerides, differences among them are the result of the number and kind of fatty acids present. A fat that predominates in fatty acids with one double bond is called monounsaturated fatty acid, a fat that predominates in fatty acids with two or more double bonds is called a polyunsaturated fat. The more unsaturated fatty acids in a fat, the more likely it is to be a liquid at room temperature. The more saturated fatty acids in a fat, the more likely it is to be a solid at room temperature.

Polyunsaturated fatty acids are further classified on the basis of their chemical structure and the position of the first double bond in relation to the methyl (CH_3) end of the molecule. Those in which the first double bond occurs between the third and fourth carbon atom are called omega-3 (ω-3 fatty acids). Fatty acids in which the first double bond is between the sixth and seventh carbon atoms are classified as belonging to the omega-6 or ω-6 family of fatty acids and those in which it occurs between the ninth and tenth carbon atoms as omega-9 or ω-9 fatty acids (Tables 3.2 and 3.3).

Table 3.2   Classification and dietary source of unsaturated fatty acids

Family	Fatty acids	No. of carbon atoms	No. of double bond	Food source
ω-3	Linolenic (18 : 3ω3)	18	3	Soybean oil, nuts
	Eicosapentaenoic acid (EPA) (20 : 5: ω3)	20	5	Fish
	Docosahexenoic acid (DHA) (22 : 6ω3)	22	6	Fish
ω-6	Linoleic (18 : 2ω6)	18	2	Vegetable oils
	Arachidonic (20 : 4ω6)	20	4	Animal tissue
ω-9	Oleic acid 18 : 1ω9	18	1	Vegetable oils

Table 3.3   Fatty acids of physiological significance

No. of C atoms and number and position of double bonds	Family	Common name	Systematic name	Occurrence
Monoenoic acids (one double bond)				
16 : 1; 9	ω-7	Palmitoleic	*cis*-9-Hexadecaenoic	In nearly all fats.
18 : 1; 9	ω-9	Oleic	*cis*-9-Octadeacenoic	Possibly the most common fatty acid in natural fats.
18 : 1; 9	ω-9	Elaidic	*trans*-9-Octadecaenoic	Hydrogenated and ruminant fats.
Dienoic acids (two double bonds)				
18 : 2; 9, 12	ω-6	Linoleic	All-*cis*-9,12-Octadecadienoic	Corn, peanut, cottonseed, soybean, and many plant oils.
Trienoic acids (three double bonds)				
18 : 3; 6, 9, 12	ω-6	γ-Linolenic	All-*cis*-6,9,12-Octadecatrienoic	Some plants, e.g. oil of evening primrose, borage oil; minor fatty acid in animals.
18 : 3; 9, 12, 15	ω-3	α-Linolenic	All-*cis*-9,12,15-Octadecatrienoic	Frequently found with linoleic acid but particularly in linseed oil.

(Contd.)

Table 3.3   (Continued)

No. of C atoms and number and position of double bonds	Family	Common name	Systematic name	Occurrence
Tetraenoic acids (four double bonds)				
20 : 4; 5, 8, 11, 14	ω-6	Arachidonic	All-*cis*-5, 8, 11, 14-Eicosatetraenoic	Found in animal fats and in peanut oil; important component of phospholipids in animals.
Pentaenoic acids (five double bonds)				
20 : 5; 5, 8, 11, 14, 17	ω-3	Timnodonic	All-cis-5, 8, 11, 14, 17-Eicosapentaenoic	Important component of fish oils, e.g. cod liver, mackerel, menhaden, salmon oils.
Hexaenoic acids (six double bonds)				
22 : 6; 4, 7, 10, 13, 16, 19	ω-3	Cervonic	All-*cis*-4, 7, 10, 13, 16, 19-Docosahexaenoic	Fish oils, phospholipids in brain.

TYPES OF FATS IN THE BODY

Lipid molecules of various types are important constituents of all cells in the body. It is convenient to divide them into three categories: structural, storage and metabolic fats.

Structural fats are those that contribute to the architecture of cells, mainly as constituents of cell membranes. In animal membranes the phosphoglycerides are the major lipids, while in plant cells the glycosyl glycerides predominate. Phosphoglycerides are also called as phospholipids.

In biological membranes, the phospholipid molecules associate together to form a bilayer that forms a continuum throughout the membrane, with the fatty acid chains pointing inwards towards each other and the polar head groups on the surfaces.

In the membranes of nervous tissue, glycolipids (where the polar moiety is sugar) are important constituents of the bilayer. Protein molecules are inserted into the phospholipid bilayer.

Lipid molecules are mobile in this layer. Cholesterol plays an important role in stabilizing the hydrophobic interactions within the membrane.

This lipid bilayer forms a barrier between the cell and its environment.

Storage fats are those that provide a long-term reserve of metabolic fuel. Triacylglycerols are the most important storage form of lipids. Storage fats contain a higher proportion of saturated or monounsaturated fatty acids and are stored in the adipose tissue.

Metabolic fats are those lipids that undergo metabolic transformation to produce a specific substance of physiological and nutritional importance.

Cholesterol is metabolized in the adrenal gland to a variety of steroid hormones and in the liver to bile acids.

Fat-soluble vitamins also participate in metabolic processes and vitamin E is stored in the membrane lipid bilayers where it serves to prevent the oxidation of the highly unsaturated membrane fatty acids.

The amphiphilic properties of phospholipids are ideal for membrane structure. They can interact with both proteins and non-polar lipids and the lipid in the bilayer provides an insulating environment for the metabolic activities of the membrane.

FUNCTIONS OF FAT

Food Fats

1. **Palatability** Foods are more palatable if they have a substantial fat content. The palatability is to be taken into account for they influence food choices. Fats contribute to palatability in two ways—by response to the texture in the mouth and by the olfactory responses of taste.

2. **Energy source** Each gram of fat, whether animal or vegetable, liquid or solid, provides 9 kcal, i.e., 2¼ times as much energy as an equal weight of either carbohydrate or protein.

3. **Satiety value** Fat tends to leave the stomach relatively slowly. It is released up until 3½ hours after being eaten, depending on size and composition of the meal. This delay in the emptying time of the stomach helps to delay onset of hunger pangs and contributes to a feeling of satiety.

4. **Carrier of fat-soluble vitamins** Dietary fat serves as a carrier of four fat-soluble vitamins: A, D, E and K. In addition, a fat level of at least 10% of total energy intake appears to be necessary for the absorption of vitamin A precursors from non-fat sources such as carrots.

5. **Essential nutrient supply** Food fats supply the essential fatty acids especially linoleic acid and cholesterol as needed to supplement the body's endogenous supply.

Body Fats

1. **Energy** A major function of fats is to provide an efficient fuel to all tissues except the central nervous system and brain, which depend on glucose.

2. **Thermal insulation** The layer of fat directly underneath the skin controls body temperature within the range necessary for life.

3. **Vital organ protection** A web-like padding of adipose tissue surrounds vital organs such as kidneys, protecting from mechanical shock and providing a supporting structure.

4. **Nerve impulse transmission** Fat layers surrounding nerve fibres provide electrical insulation and transmit nerve impulses.

5. **Tissue membrane structure** Fat serves as a vital constituent of cell membrane structure, helping in the transport of nutrient materials and metabolites across cell membranes.

6. **Cell metabolism** Lipoproteins, a combination of fat and proteins, carry fat in the blood to all the cells.

7. **Essential precursor substances** Fat supplies necessary components such as fatty acids and cholesterol for synthesis of many materials required for metabolic function and tissue integrity.

Essential Fatty Acids (EFA)

These are fatty acids which are a necessity in the diet. These are necessary for body functions and metabolism and cannot be manufactured by the body and must therefore be supplied in the diet. Linoleic, linolenic and arachidonic are considered as essential. These serve important functions in the body.

1. They strengthen capillary and cell membrane structure, which helps prevent an increase in skin and membrane permeability. A deficiency of linoleic acid leads to a breakdown in skin and other tissue membrane integrity.

2. They combine with cholesterol for its transport in the blood.

3. They help to lower serum cholesterol. Linoleic acid plays a key role in the transport and metabolism of cholesterol.

4. They prolong blood clotting time and increase fibrinolytic activity.

5. They help to form a group of physiologically and pharmacologically active compounds known as prostaglandins. These substances are synthesized in the body from the essential fatty acids especially arachidonic acid and its precursor, linoleic acid.

During EFA deficiency, changes occur in the properties of membranes, the permeability to water and small molecules for example, which can be correlated with changes in the fatty acid composition of the membrane. β-oxidation and oxidative phosphorylation are less efficient. It seems that the stability and integrity of the membrane and its ability to provide an environment for the efficient functioning of the enzymes, receptors and other proteins embedded in the lipid bilayer, can only be supported by the presence of lipids with a certain pattern of polyunsaturated fatty acids.

Eicosanoids

Arachidonic acid and other C-20 and C-22 polyunsaturated fatty acids of the n-3 and n-6 families can be metabolized to a range of compounds that exert influence on many physiological activities. These include:

a) Contraction of smooth muscle,

b) Inhibition or stimulation of the adhesion of blood platelets and

c) Constriction or dilation of blood vessels with related influence on blood pressure.

The eicosanoids inlcude prostaglandins, prostacyclins, thromboxanes and leukotrienes.

Prostacyclins and thromboxanes have opposite physiological effects. Prostacyclins formed in arterial walls are among the most powerful known inhibitors of platelet aggregation. They relax arterial walls and promote a lowering of blood pressure. Thromboxanes, found in platelets, stimulate platelets to aggregate, are important for wound healing, contract the arterial wall and promote an increase in blood pressure. The balance between these activities is important in maintaining normal vascular function.

The amounts and types of n-6 and n-3 fatty acids in the diet influences the spectrum of eicosanoids produced. If fish oils in which n-3 fatty acids predominate in diets, it results in changes in plasma and platelet fatty acid profiles. There is a reduction in the formation of thromboxane A_2 and an increase in the formation of thromboxane A_3 by platelets. Blood platelets are stimulated to aggregate less strongly by the 3-series of thromboxanes than those of the 2-series, and thrombotic tendency is reduced. Moreover, the n-3 PUFA also tend to directly inhibit the cyclo-oxygenase enzyme responsible for eicosanoid biosynthesis.

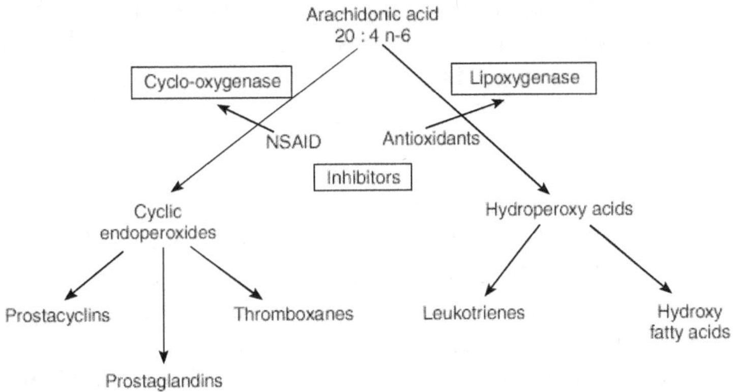

Figure 3.1   The two major pathways of arachidonic acid metabolism to eicosanoids. NSAID = non-steroidal anti-inflammatory drugs like aspirin.

PUFA are also metabolized to another type of eicosanoid known as leukotrienes. Arachidonic acid (n-6) is metabolized to leukotriene B_4 which has considerable inflammatory action while eicosapentaenoic acid (n-3) is metabolized to leukotriene B_5 which is less inflammatory (Figure 3.1).

DIGESTION OF FAT

The basic fat fuels—animal and plant fats (triglycerides)—are converted to a refined fuel form of fat, the fatty acids, that the cells can burn for energy.

Mouth

No fat digestion takes place in the mouth. In this portion of the GI tract, fat is simply broken down into smaller particles through chewing and moistened for passage into the stomach with the general food mass.

Stomach

General peristalsis continues the mechanical mixing of fats with stomach contents. Very little digestion of fats takes place in the stomach. No significant amount of enzymes specific for fats is present in the gastric secretions except a gastric lipase which acts on emulsified butterfat.

Small intestine

Bile from liver and gall bladder   The presence of fats in the duodenum stimulates the secretion of cholecystokinin, a local hormone from the glands in the intestinal walls. Cholecystokinin causes contraction of the gall bladder, relaxation of the sphincter and subsequent secretion of bile into the intestine via the common bile duct. The liver produces a large amount of dilute bile, then the gall bladder concentrates and stores it. It functions as an emulsifier in the following ways.

1. It breaks the fat into small particles, or globules, which greatly enlarges the surface area available for action of the enzyme

2. It serves to lower the surface tension of the finely dispersed and suspended fat globules which allows the enzymes to penetrate more easily.

3. It provides an alkaline medium for the action of lipase.

Enzyme from the pancreas   Pancreatic juice contains an enzyme for fat and one for cholesterol. First, pancreatic lipase, a powerful fat enzyme breaks off one fatty acid at a time from the glycerol base of fats. The final products of fat digestion are fatty acids, diglycerides, monoglycerides and glycerol. Some remaining fat may pass into the large intestine for faecal elimination. Second, the enzyme cholesterol esterase, acts on cholesterol to form cholesterol esters by combining cholesterol and fatty acids in preparing free cholesterol for absorption.

Enzyme from small intestine   The small intestine secretes an enzyme in the intestinal juice called lecithinase. It acts on lecithin to break it down into its components for absorption.

ABSORPTION

Fats require a solvent carrier. Absorption takes place in three steps.

Step I

Bile combines with products of fat digestion in a micellar bile–fat complex and carries fat into the intestinal wall.

Step II

Absorption within the intestinal wall   Bile separates again from the fat complex and is returned in circulation to accomplish its task over and over again. Two important actions on the fat products occur inside the intestinal wall:

1. *Enteric lipase action* An enteric lipase within the cells of the intestinal wall completes the digestion of remaining glycerides and

2. *Triglyceride synthesis* With the resulting fatty acids and glycerol, new human triglycerides are formed as body fats.

Step III

These newly formed human fats, triglycerides, and other fat materials present are combined with a small amount of protein to form lipoproteins called chylomicrons. These fat packages in a milk-like liquid called chyle cross the cell membrane intact into lymphatic system and then into the portal blood. Here a final fat-clearing enzyme, lipoprotein lipase, helps clean the large load of dietary fat from circulation. In the liver, fat is converted to other lipoproteins for transport to the cells for energy and other structural functions.

METABOLISM OF LIPIDS

Fatty acids are both oxidized to acetyl-CoA and synthesized from acetyl-CoA. The term "free fatty acid" (FFA) refers to fatty acids that are in unesterified state. In plasma, the free fatty acid of longer-chain fatty acids are combined with albumin, and in the cell they are attached to a fatty-acid-binding protein or z-protein. Shorter-chain fatty acids are more water-soluble and exist as the un-ionized acid or as a fatty acid anion.

Activation of fatty acids The fatty acids must first be converted in a reaction with ATP to an active intermediate before they will react with the enzymes responsible for their further metabolism. The activated long-chain fatty acids will not penetrate the mitochondria and become oxidized unless it forms acylcarnitine. An enzyme carnitine palmitoyl transferase I is associated with the outer side of the inner mitochondrial membrane and converts long-chain acyl groups to acylcarnitine, which are able to penetrate mitochondria and gain access to the β-oxidation system of enzymes (Figure 3.2).

Long-chain acyl-CoA cannot pass through the inner mitochondrial membrane, but its metabolic product acylcarnitine formed by the action of acylcarnitine palmitoyl transferase I can. Carnitine–acylcarnitine translocase acts as a membrane carnitine exchange transporter. Acylcarnitine is transported in, coupled with the transport out of one molecule of carnitine. The acylcarnitine then reacts with CoA, catalysed by carnitine palmitoyl transferase II, attached to the inside of the inner membrane. Acyl-CoA is reformed in the mitochondrial matrix, and carnitine is liberated.

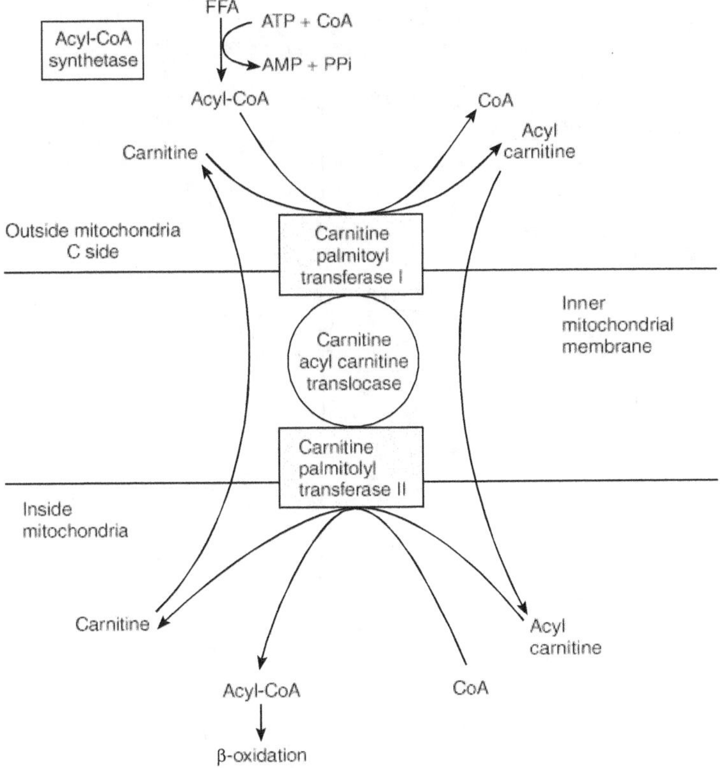

Figure 3.2   Role of carnitine in the transport of long-chain fatty acids through the inner mitochondrial membrane

β-oxidation

In β-oxidation, 2 carbons are cleaved at a time from acyl-CoA molecules, starting at the carboxyl end. The chain is broken between the α(2) and β(3) carbon atoms, hence the name β-oxidation. The 2-carbon units formed are acetyl-CoA, thus palmitoyl-CoA forms 8 acetyl-CoA residues (Figure 3.3).

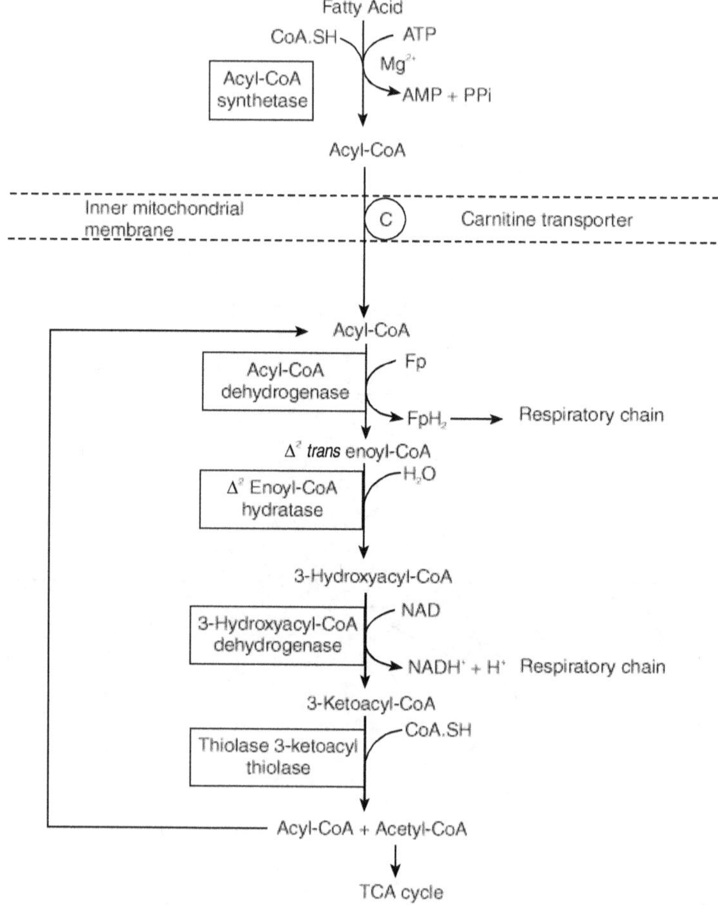

Figure 3.3   β-oxidation of fatty acids

Long-chain acyl-CoA is cycled through the reactions, acetyl-CoA being split off each cycle by thiolase. When the acyl radical is only four carbon atoms in length, 2-acetyl-CoA molecules are formed in the final reaction.

Biosynthesis of Saturated Fatty Acids

A mitochondrial system for fatty acid synthesis involving some modification of the β-oxidation sequence is responsible only for the elongation of existing fatty acids of moderate chain length, whereas a radically different and highly active extramitochondrial system is responsible for the complete synthesis of palmitate from acetyl-CoA. An active system for chain elongation is also present in liver endoplasmic reticulum.

Extramitochondrial system for lipogenesis is found in the soluble fraction of many tissues, including, liver, kidney, brain, lung, mammary gland and adipose tissue. Its cofactor requirements include NADPH, ATP, Mn^{2+} and HCO_3^- (as a source of CO_2), acetyl-CoA is the substrate, and free palmitate in the end product.

Bicarbonate as a source of CO_2 is required in the initial reaction for the carboxylation of acetyl-CoA to malonyl-CoA in the presence of ATP and acetyl-CoA carboxylase (Figure 3.4). Acetyl-CoA carboxylase has a requirement for the vitamin biotin. The enzyme contains a variable number of identical subunits, each monomer containing biotin, biotin carboxylase, biotin carboxyl carrier protein, and transcarboxylase. It is therefore a multienzyme protein. It may not be subdivided without loss of activity. The acyl carrier protein contains the vitamin pantothenic acid in the form of 4'-hosphopantetheine. The fatty acid synthase complex is a dimer and each monomer is identical.

The complex is a dimer of 2 identical polypeptide monomers each consisting of 6 separate enzyme activities and the acyl carrier protein (ACP) (Figure 3.5). The SH of the 4'-phospho-pantetheine of one monomer is in close proximity to the —SH

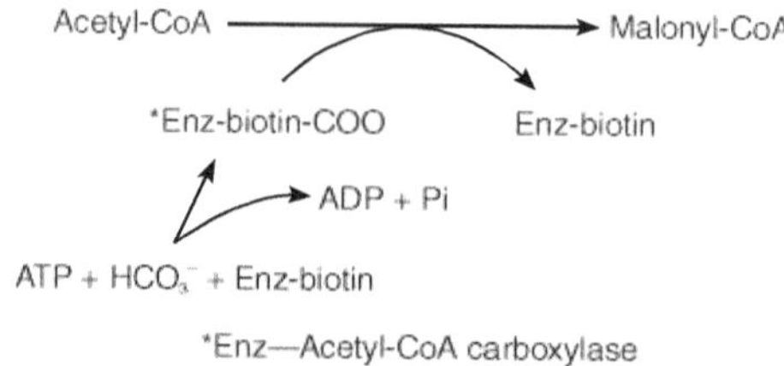

Figure 3.4   Production of malonyl-CoA

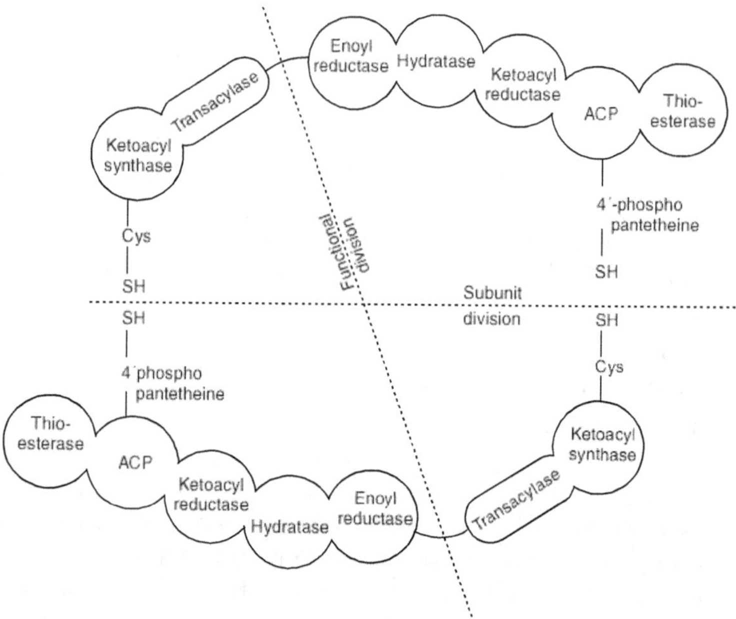

Figure 3.5   Fatty acid synthase multienzyme complex

of the cysteine residue of the ketoacyl synthase of the other monomer, suggesting a "head-to-tail" arrangement of the 2 monomers. The actual functional unit consists of one-half of a monomer interacting with the complementary half of the other. Thus 2 acyl chains are produced simultaneously (Figure 3.6).

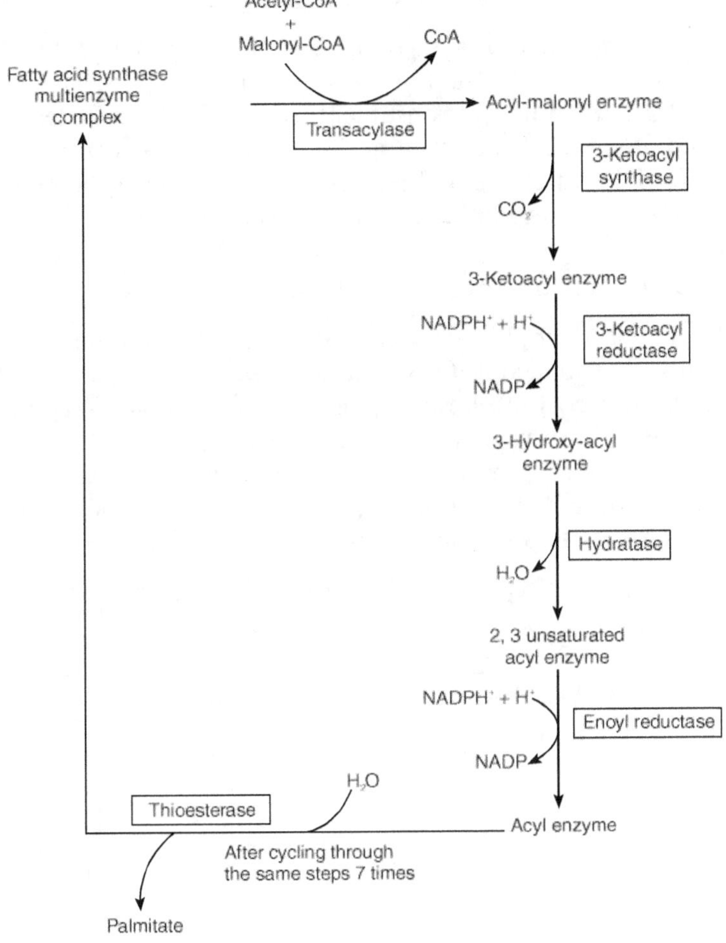

Figure 3.6 Biosynthesis of fatty acids

TRANSPORT

Absorbed lipids are transported in water-soluble form from the small intestine to other tissues. The FA with chain lengths shorter than 14C atoms are bound to albumin and preferentially transported directly to the liver by way of the portal vein.

Principles of Intravascular Transport

Of the three main classes of lipids (TG, PL and CH), the TG are used primarily as a fuel or are stored in fat depots. The PL and CH are the principal constituents of plasma and biomembranes and also participate in intracellular transport as constituents of micelles and emulsified particles. They are also secreted with bile salts as mixed micelles into the canicular bile.

Chylomicrons, VLDL and LDL

The long-chain TG are transported from the intestinal mucosal cells as chylomicrons (TG synthesized in the enterocyte acquire a stabilizing coat of phospholipids and apolipoproteins, apo A and apo B. These chylomicrons, large spherical particles 75–600 nm in diameter, are secreted into the lymphatic vessels and pass via the thoracic duct into the jugular vein. In the bloodstream they acquire apolipoproteins C and E.

Table 3.4 Composition and characteristics of the human plasma lipoproteins

Chylomicrons	VLDL	LDL	HDL
2	7	20	50
83	50	10	8
8	22	48	20
7	20	22	22
> 70	30–90	18–22	5–12

Lipids are carried in the blood, bound to protein in complex macromolecules called lipoproteins. Four classes of lipoproteins are characterized by their density (and electrophoretic mobility). Each contains triglycerides (TG) phospholipids (PL) and cholesterol (CH) bound to a protein carrier (apoprotein) in large molecular complexes. In each of these four classes, the proportion of these components in the complexes varies slightly but differs markedly from the other classes (*see* Table 3.4 for composition).

Apoproteins   These are protein carriers made up of several units of relatively small proteins. The most prevalent apoproteins are called A, B and C. These are synthesized by liver and mucosal cells of the small intestine.

Chylomicrons   These are the main vehicles for carrying triglycerides from the alimentary canal to the liver and they also carry phospholipids, cholesterol, and fat-soluble vitamins in relatively small amounts. They are cleared from the circulation usually 2–4 hours after a meal by lipoprotein lipases on the surface of endothelial cells of the capillary bed of muscle, adipose tissue and liver. These enzymes hydrolyse the triglycerides and free fatty acid and are taken up by the cells. The chylomicrons are progressively reduced in size to small particles known as remnants. From these, LDL may be formed in the plasma or they may be taken up by the liver and metabolized.

Very Low Density Lipoprotein (VLDL)   These are the carriers of endogenous triglycerides, from the liver to the peripheral tissues. After assembly, they enter the bloodstream. Like the chylomicrons, they are then subjected to the action of lipoprotein lipase in the extrahepatic (outside the liver) capillary bed. Fatty acids and monoglycerides are liberated and absorbed onto the cell. The protein portion of the complex with a little remaining lipid stays in circulation where they are called as remnants and from these the LDL are formed subsequently. The plasma concentration of VLDL changes little and unlike that of chylomicron is not closely related to the ingestion of fats. Fasting plasma triglyceride concentration reflects VLDL concentration after an overnight fast. The normal values range from 25 to 150 g per 100 ml.

Low density lipoproteins (LDL) and cholesterol metabolism   Cholesterol is carried in the blood mainly on the LDL and in the esterified form. Cholesterol in the food and also in the biliary secretion is absorbed in the small intestine and carried to the liver in the chylomicron. Then, together with any synthesized in the liver it enters the circulation as LDL. Cholesterol, however, is only a small portion of the total lipid in chylomicrons and VLDL functions as the main carrier of triglycerides and the amount of triglycerides absorbed in the gut and synthesized in the liver is much greater than that of cholesterol.

LDL   LDL contains much more cholesterol than triglyceride and is formed in the plasma from the remnants and also from cholesteryl esters made available from phospholipids through the action of the enzyme LCAT (lecithin cholesterol acyl transferase). This enzyme is activated by another lipoprotein, HDL, formed in the liver. Free cholesterol may also leave the cells and then is esterified and taken up into LDL through the action of HDL.

LDL is taken up by specific receptors on the cell membrane. It is then broken down and the cholesterol used to restructure the membrane which are continuously renewed. LDL appears to function as a mobile reserve of cholesterol from which cell can draw through their specific receptors at a rate determined by the need for structural renewal. LDL is also taken up by the cells by a route independent of the receptors and is called as the scavenger pathway.

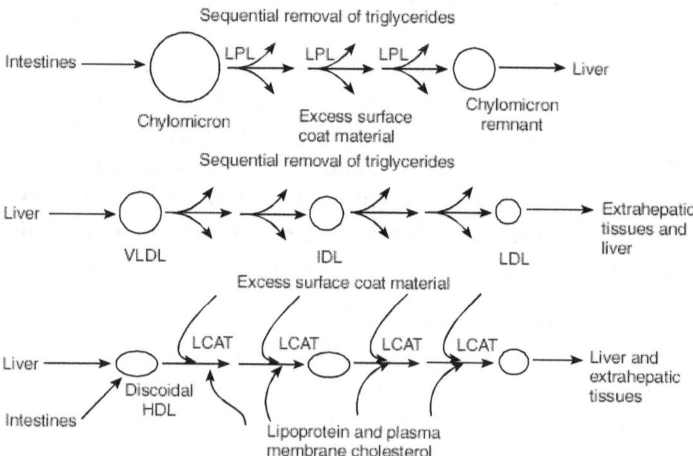

Figure 3.7   A schematic representation of lipid transport processes. LPL—lipoprotein lipase, IDL—Intermediary lipoprotein, LCAT—Lecithin cholesterol acyl transferase.

The cholesterol content of an adult man is about 140 g. The turnover of this mobile pool in healthy subjects is about 1.5 g daily and indirect evidence indicates that about two-thirds of this goes through specific receptor sites and one-third through the scavenger pathway. Plasma cholesterol in healthy people ranges from 3.6–7.8 mmol/litre (140–300 mg/100 ml) and reflects the plasma LDL concentration (Figure 3.7).

ROLE OF LIPIDS IN CORONARY HEART DISEASE (CHD)

CHD is a condition in which main arteries supplying blood to the heart are no longer able to supply sufficient blood and oxygen to the heart muscle (myocardium) which may then quickly die. The main cause of reduced blood flow is the accumulation of deposits (plaques) in the linings (intima) of the arteries, a disease known as **Atherosclerosis**.

Lipid hypothesis The lipid hypothesis promotes the concept that elevation in blood lipid induces atherosclerosis or that, they are responsible for the process of atherogenesis (Figure 3.8).

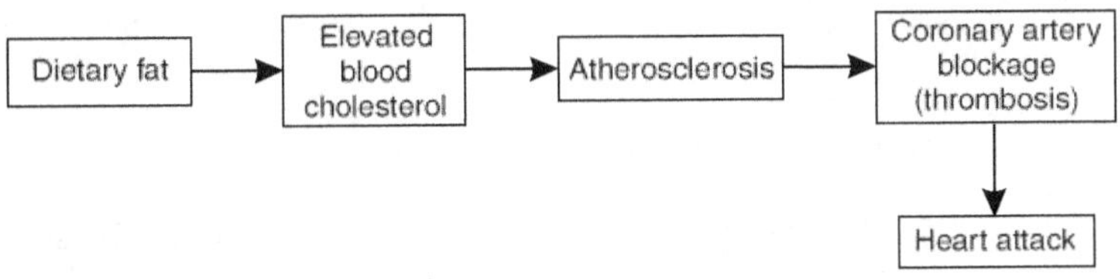

Figure 3.8   Process of elevation of cholesterol and relation to heart disease

This implies that atherosclerosis might be prevented by appropriate dietary modification.

Lipids are also thought to be involved in the development of atherosclerosis in two main areas:

1. First, lipids are involved in the initiation of lesion.

2. Secondly, they are directly involved in exacerbating the disease.

Saturated Fatty Acids and CHD

Dietary saturated fatty acids tend to elevate cholesterol level whether they are of vegetable or animal origin [coconut oil, cocoa butter, palm oil, and lard].

It has been observed that dietary saturated fats are twice as powerful in increasing plasma cholesterol as is PUFA in decreasing plasma cholesterol (Figure 3.9).

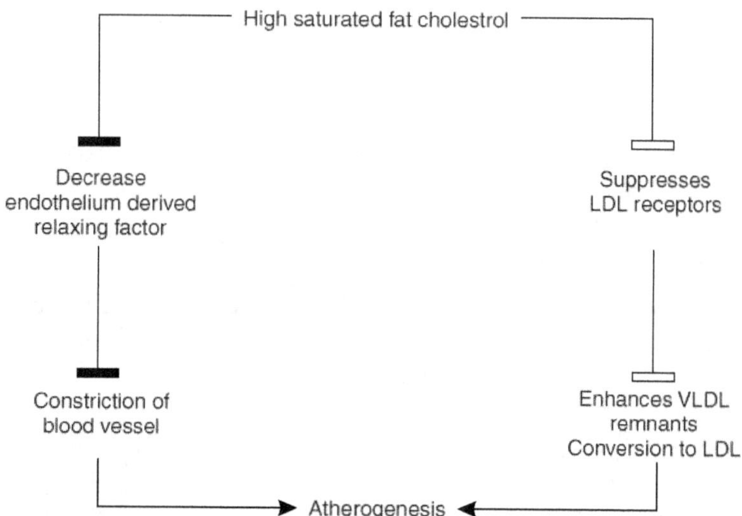

Figure 3.9   Saturated fat and cholesterol in causing atherogenesis

Cholesterol and CHD

The risk of atherosclerosis increases linearly as cholesterol level increases from about 180–200 mg/dl. It therefore may be desirable to control serum cholesterol at these levels throughout adult life.

Dietary cholesterol intake influences plasma cholesterol level profoundly.

As intake is increased from 0–300 mg, the increase in plasma cholesterol is about 9 mg/100 ml up to 300 mg/1000 kcal.

Thereafter the increase in plasma cholesterol is 2 mg/100 mg/1000 kcal.

Lipoproteins and CHD

Cholesterol and TGs are transported in plasma as lipoprotein. Diets high in saturated fats and cholesterol produce alterations in plasma lipoproteins and these alterations cause certain lipoproteins to deliver cholesterol to the arterial wall. Atherosclerosis occurs when the influx of cholesterol into the artery exceeds the efflux of cholesterol from tissues.

Chylomicron

It is not clear whether patients with chylomicronaemia are at increased risk for atherosclerosis. But remnants of chylomicron may be atherogenic.

VLDL

VLDL is produced in the liver and intestine. VLDL is the main transporter of endogenous TG which is synthesized when there is excessive intake of calories. Elevation of plasma TG is common in patients with CHD but the role of plasma TG as an independent risk factor is in dispute.

Beta VLDL are found to be atherogenic. Beta VLDL can be internalized by the extrahepatic tissues including the smooth muscle cells of arterial walls. In particular, macrophage monocytes in the arterial wall internalize beta VLDL and accumulate tremendous amount of cholesterol and then they resemble the **foam cells** seen in atherogenesis.

LDL

Removal of TG from VLDL and chylomicron by lipase generates LDL. LDL is the main transporter of cholesterol in plasma. LDL is about 50% cholesterol. LDL is taken up by the peripheral tissues and liver by the specific cell-surface lipoprotein receptors.

The absence or malfunctioning of these receptors is seen in familial hypercholesterolaemia, and with increased SFA intake.

LDL is the villain in atherogenesis. It is the most atherogenic of all the lipoproteins. Increased plasma level of LDL are correlated strongly with manifestation of atherosclerosis.

1. It may be that chemically modified LDL has longer half-life.

2. Increased size of LDL may promote a greater accumulation of cholesterol in smooth muscle cells than normal LDL.

3. Another possible role of LDL may be stimulation of cell growth and proliferation, characteristic of atherogenic process.

4. Modified LDL interacts with macrophages which have receptors called scavenger receptors, and are gobbled by them.

In course of time, the atherosclerotic plaque becomes full of these lipid-filled macrophages which are called foam cells.

The dietary factors that decrease circulating cholesterol include reduced intake of saturated fat and cholesterol, increased intake of PUFA, substitutes like soyabean protein for meat protein, and complex carbohydrates for simple carbohydrates.

HDL

HDL has been called the good lipoprotein, since HDL cholesterol levels are inversely related to the risk of atherosclerosis. HDL carries our reverse cholesterol transport, taking cholesterol and delivering it to the liver for catabolism to bile acids and resultant loss from the body. Hence HDL is anti-atherogenic, with an inverse relationship between HDL level and coronary artery disease.

There are a variety of dietary effects on HDL. An increase in simple sugars or a high CHO diet for short periods result in a decrease in HDL cholesterol and apoprotein, presumably by accelerating removal of HDL cholesterol. These diets also increase VLDL and TGs. Increase in PUFA or decrease in saturated fat results in increase of HDL cholesterol and apoprotein. These diets also decrease LDL cholesterol. Diets high in fat or cholesterol on the other hand increase LDL cholesterol levels.

PUFA and CHD

Omega-6 fatty acids and plasma lipid The hypocholesterolaemic effect of PUFA present in vegetable oils has been appreciated for 30 years.

Kinsell and colleagues (1952) suggested that it was not simply the amount of fat but the type of fat in the diet (animal or vegetable) that affected the plasma cholesterol level.

The Iodine Number (total unsaturation) of the oil rather than its source was correlated with the degree of cholesterol lowering.

Keys and colleagues, and Hegsted and colleagues studied that gram by gram, saturated fatty acids raised cholesterol level about twice as much as PUFAs lower them.

The proposed mechanism by which polyunsaturated omega-6 fatty acids exert their hypolipidaemic effect can be given under the following steps.

1. Decreased cholesterol level absorption.

2. Increased faecal excretion of steroids.

3. Decreased cholesterol synthesis.

4. Transfer of cholesterol from plasma to tissues.

5. Changes in cholesterol to protein ratio in LDL.

6. Changes in rate of synthesis or catabolism of individual lipoprotein.

In addition, changes in fatty acid composition of membrane lipid may affect lipoprotein metabolism in 2 ways:

1. Changes in lipoprotein particles may make them a better substrate for catabolic enzymes or for binding to cellular receptors.

2. Changes in cellular membrane may increase reactivity towards the lipoprotein substrate.

OMEGA-3 FATTY ACID AND PLASMA LIPID LEVEL

Fish oils were found to be hypocholesterolaemic but contain large amounts of cholesterol 300–500 mg/100 g while vegetable oils contain none. Findings have therefore implied that fish oil might have been even more hypocholesterolaemic if they had not contained cholesterol.

Secondly the intake of omega-6 fatty acid was invariably greater than intake of omega-3 fatty acid to get similar results or greater reduction in plasma cholesterol level occurred with omega-3 fatty acid.

Thus gram for gram, omega-3 fatty acids are considerably more hypocholesterolaemic than linoleic acid.

It was shown that approximately 4 g/day of omega-3 fatty acid from 20 ml of cod liver oil produced greater cholesterol lowering than did 16 g of linoleic acid.

Dramatic reduction in the concentration of plasma TGs and VLDL were observed in both normolipidaemic and hyperlipidaemic subjects fed diets from supplemented fish oils.

It was further demonstrated that dietary omega-3 fatty acid significantly decreased plasma LDL [16%], VLDL [50%] and did not change the concentration of HDL and in some studies has actually raised HDL cholesterol levels.

The suggested mechanisms for omega-3 fatty acid-induced hypocholesterolaemia include:

1. Increased faecal steroid excretion.

2. Changes in the rate of synthesis and catabolism of VLDL and LDL.

Reduction in concentration of TGs and VLDL may result from the lower rate of synthesis of the TG or the apoprotein.

ROLE OF LIPIDS IN CORONARY THROMBOSIS

Lipids are involved in the biology of thrombotic process and there is growing evidence that dietary fats are able to influence the tendency of a thrombus to form.

The surface of the blood vessels are subjected to continual wear and tear and the body has evolved a complex repair system. If the repair process were not under strict control, there could be production of some of the eicosanoids which cause excessive clotting of the blood and accumulation of material on the wall of the artery which cause unwanted and dangerous clots to be formed. The formation of clots in thrombosis may be regarded as aspects of the failure of control mechanism. Overproduction of these eicosanoids can be regulated by EPA and DHA found in fish oil. EPA and DHA act to decrease the stickiness of blood platelets. It is this stickiness that causes platelets to adhere to each other and to form a clot. Thus EPA and DHA reduce the possibility of a heart attack.

REVIEW QUESTIONS

1. Give the functions of lipids.

2. Discuss the functions of EFA.

3. How are lipids implicated in the development of heart diseases?

4. What are n-3 fatty acids and how do they prevent atherosclerosis?

5. Write a note on absorption of fats.

CRITICAL THINKING QUESTIONS

1. Why is fat an absolute necessity in our diet?

2. Is blood cholesterol test necessary at age 40 or below?

4

PROTEINS

Did You Know?

- An insufficient protein intake can stunt a child's growth. Protein is particularly important for building new tissue during periods of rapid growth.

- The hormone insulin is a protein. It contains only amino acids.

- The quality of protein can be measured by its biological value—nitrogen retention divided by nitrogen absorption. Biological value represents the body's ability to retain the protein absorbed.

- Milk provides higher quality proteins than most other foods. Milk proteins provide one of the highest possible biological values from foods. An egg white has the very best biological value.

- Animal protein sources often contain high amounts of saturated fat. For this reason, trimming meats of fat and broiling them is a good idea.

- Lack of energy can be a symptom of severe protein deficiency. Although lack of energy can be caused by other things, it is a symptom of severe protein deficiency.

- Marasmus is a disease caused by starvation and can be seen in large cities of impoverished countries. Starvation in infancy leads to marasmus, which means "to waste away".

- Water-packed tuna is almost pure protein. It is 85% protein.

- Fruits contain very little protein. These contain mostly carbohydrate and water.

The term protein, meaning "to take first place" was first introduced by Mulder, a Dutch chemist, in 1838. He defined protein as a nitrogen-containing constituent of food and believed it to be of such importance in the functioning of the body that, without it life would be impossible.

Proteins are extremely complex nitrogenous organic compounds in which amino acids are the units of structure. They contain the elements carbon, hydrogen, oxygen and nitrogen and some contain sulphur. Most proteins also contain phosphorus and some specialized proteins contain very small amounts of iron, copper and other inorganic elements. The presence of nitrogen distinguishes protein from carbohydrate and fat. Proteins contain an average of 16% nitrogen. The large protein molecules form colloidal solutions that do not readily diffuse through membranes.

STRUCTURE OF PROTEINS

Proteins are macromolecules consisting of long chains of amino acid subunits. The general formula for an amino acid is,

$$R-CH-COOH$$
$$|$$
$$NH_2$$

In the protein molecule, the amino acids are joined together by peptide bonds, which result from the elimination of water between the carboxyl group of one amino acid and the amino group of the next. In biological systems, the chains so formed might be anything from two amino acid units (dipeptide) to thousands of units long, corresponding to molecular weights ranging from hundreds to hundreds of thousands of Dalton.

The polypeptide chains do not exist as long straight chains, nor do they curl up into random shapes but instead fold into a definite three-dimensional structure. Some proteins are globular in shape, some are rods, while some are composed of several separate peptide chains held together by ionic or covalent links.

The most important aspect of proteins from a nutritional point of view is their amino acid composition, but the protein's structure influences its nutritional availability. Some proteins are highly insoluble and are resistant to digestion (e.g. hair) while some others are resistant to attack by the hydrolytic enzymes (e.g. mucins secreted by the intestines).

AMINO ACIDS

The difference between the twenty or so amino acids found in proteins lies in the structure of the side chain designated as R. The amino acids not only vary in size but the side groups also carry different charges, some are hydrophobic while others are hydrophilic.

These side chains determine the protein structure and many other aspects of protein function. Attraction between positive and negative charges pull different parts of the molecule together. Hydrophobic groups tend to cluster together in the centre of globular proteins, while hydrophilic groups remain in contact with water on the periphery.

Essential Amino Acids

Eight of the amino acids are marked as "nutritionally essential" as the carbon skeletons of these amino acids cannot be synthesized from simpler molecules and therefore must be provided in the diet. These are valine, leucine, isoleucine, phenylalanine, tryptophan, threonine, methionine and lysine.

Histidine and arginine are classified as **semi-essential amino acids**. These are so considered because the rate of synthesis in the body is inadequate to support growth; therefore these are essential for children. Recent studies indicate that some histidine may also be required by adults.

Non-essential Amino Acids

The remaining amino acids, i.e., glutamic acid, aspartic acid, alanine, proline and hydroxy proline are those amino acids which can be synthesized from the existing carbon skeleton and hence are considered as not essential in the diet.

Proteins are classified in a number of ways based on their physical properties, chemical properties, physical shape and their nutritional constituents. The classification of proteins based on their amino acid composition is as follows:

1. *Complete proteins* A protein which contains a sufficient quantity of essential amino acids to maintain body tissues and promotes a normal rate of growth is considered as a complete protein, e.g. egg, milk and meat proteins.

2. *Partially complete proteins* These will maintain life but they lack sufficient amounts of the essential amino acids necessary for growth. Examples of these are cereals, legumes, nuts and seeds where one or more essential amino acid is limiting.

3. *Totally incomplete proteins* These proteins are incapable of replacing or building new tissues and hence cannot support life. Zein found in corn, and gelatin are classic examples of proteins that are incapable of even permitting life to continue.

FUNCTIONS

Proteins function in three main ways—building tissue, performing various specific additional roles and sometimes providing energy.

Growth and maintenance of tissue Dietary protein furnishes amino acids of appropriate numbers and types for efficient synthesis of specific cellular tissue proteins. It also provides sufficient amino groups to form the non-essential amino acids. Some tissues call for larger amounts of specific amino acids. The hair, skin and nails for example require larger amounts of sulphur-containing amino acids.

The protein of the body is in a constant dynamic state. It is alternately broken down and resynthesized, with about 3% of total body protein being turned over each day. The wall of the intestine, which is replaced every 4 to 6 days requires the synthesis of 70 g of protein a day. However, the body conserves protein and reuses the amino acids from the breakdown of tissues to build more of the same or other tissues.

Specific physiological roles Many proteins perform specific physiological and metabolic roles.

1. Hormones like insulin, gastrin, growth hormone, etc. are proteins.

2. All the enzymes that are involved in digestion and metabolism are proteins.

3. Haemoglobin responsible for carrying both oxygen and carbon dioxide in the RBC is a protein substance.

4. Many of the blood clotting factors are proteins.

5. The photoreceptors in the eye responsible for vision are proteins.

6. The catecholamines, adrenaline or epinephrine is produced from tyrosine.

7. The amino acid tryptophan serves as a precursor for the vitamin niacin and for serotonin which is a neurotransmitter.

8. Glutamic acid, cysteine and glycine are components of glutathione, which functions in cellular oxidation–reduction reactions.

9. Lysine is the parent substance of carnitine.

10. Tyrosine is the precursor of melanin.

11. Methionine is an agent in the formation of choline which is a precursor of acetylcholine, a neuro-transmitter.

12.   Glycine is involved in detoxifying mechanisms. Many drug metabolites and other compounds with carboxyl groups are excreted in the urine as glycine conjugates.

13.   Glycine is incorporated into creatine and the pyrrole ring of haem and in purines.

Regulation of water balance Fluid in the body is contained in two different compartments—the intracellular compartment and the extracellular which includes intercellular compartment and intravascular. These compartments are separated from one another by cell membranes. The distribution of fluid among them must be kept in balance. This balance is achieved through a complex system of controls involving both protein and electrolytes, primarily sodium and potassium. Protein in the blood is too large to pass out of the bloodstream and hence exerts an osmotic pressure, which draws fluid from the intercellular compartment back into the blood. When blood proteins are low, the osmotic pressure of the protein pulling fluid back into circulation is not as strong as the hydrostatic pressure pushing it out of the bloodstream. This results in an accumulation of fluid in the tissues that results in oedema and is an early sign of protein deficiency.

Maintenance of body pH Proteins serve as buffers and help to neutralize acids or bases. For example histidine in haemoglobin functions as a very efficient buffer.

Antibody formation The body's ability to produce antibodies is a very important function of proteins which helps fight infections.

Detoxification The ability to detoxify, or remove poisonous material from the body is controlled by enzymes located primarily in the liver. In protein depletion the ability to counteract the toxic effect of chemicals is reduced. Methionine protects the liver from damage by poisons such as carbon tetrachloride, arsenic or chloroform.

Transport of nutrients　Proteins function as carriers of nutrients from the intestine across the intestinal wall to the blood, from the blood to the tissues and across the cell membrane into the cell. A classical example is lipoproteins which help transport lipids or retinol-binding protein which helps transport retinol.

Source of energy　Proteins in excess of needs for other functions is used as a source of energy or stored as fat.

DIGESTION

All dietary proteins that consist of complex units of amino acids are too large to pass through the intestinal wall. They must be broken down to simpler units consisting of dipeptides and amino acids. This involves the hydrolysis of the peptide linkages of a protein.

The food is chewed and mixed with saliva in the mouth and enters the stomach ready for digestion.

In the stomach, hydrochloric acid is released which decreases the pH of the stomach contents and converts inactive pepsinogen to active pepsin. Pepsin breaks down the protein of the diet into smaller peptide units while hydrochloric acid denatures the protein so that further digestion can take place in the intestines.

The partially digested food then enters the small intestines, where bicarbonate from the pancreatic juice neutralizes acidity in the food mass to allow intestinal enzymes to digest the protein further. The enzymes secreted by the pancreas are trypsin, chymotrypsin, elastase, collagenase and carboxypeptidase. These enzymes attack specific peptide bonds on the peptide chain and break it down further into smaller peptide units.

The final digestion of protein to individual amino acids or to dipeptides and tripeptides is accomplished by enzymes available from the mucosal cells lining the small intestines, which act either within the mucosal membrane or within the mucosal cells. Final release of amino acids from the peptides occurs within the mucosal cell.

ABSORPTION

The amino acids, dipeptides and tripeptides are absorbed. Absorption of amino acids is dependent on the carrier capable of transporting one of five particular amino acids.

Once the amino acids reach the liver through the portal vein, some are synthesized into plasma proteins. The remainder is released into the general circulation to be transported to the individual cells of the body to make the unique proteins that the cells need.

METABOLISM

When the energy intake is adequate, the amino acids derived from dietary proteins are used first for synthesizing body proteins. However, those amino acids in excess of needs for growth and maintenance undergo deamination and lose their amino group and the remaining carbon skeleton of amino acids enters the TCA cycle at different points depending on the structure or is used to synthesize other non-essential amino acids. The amino portion is converted to urea and excreted by the kidneys.

Some of the amino acids are glucogenic (so called for they serve as sources of glucose after deamination) or are ketogenic since they produce ketones.

FACTORS AFFECTING PROTEIN UTILIZATION

Calorie intake   When the calorie intake is inadequate, protein is deaminated and used for energy.

Immobility   The ability to synthesize protein is greatly reduced among people who are immobile. Older people who are bedridden lose protein mass even when dietary protein and calorie intake seem adequate.

Injury   An increase in nitrogen loss occurs after injury. High protein intake either before or after injury does not prevent this loss.

Emotional stress   The emotional stresses such as fear, anxiety or anger increase the secretion of epinephrine from the adrenal gland, which in turn causes a series of changes that result in the loss of nitrogen.

Amino acid imbalance   The term amino acid imbalance is applied to an amino acid pattern containing either excessive or inadequate amounts of certain essential amino acids. The mixture causes a depression in growth which is completely prevented by a small supplement of the limiting amino acids. For example, the growth of albino rats, fed a protein limiting in methionine and supplemented with other essential amino acids, is depressed. This aggravates methionine deficiency and produces imbalance with great depression of growth. Addition of methionine along with other amino acids overcomes the growth depression.

Amino acid antagonism   The term antagonism is applied to those changes in the amino acid pattern of a diet that cause a growth depression which is largely or completely prevented by the addition of small amounts of a structurally similar amino acid or amino acids not originally limiting in the diet.

For example, when leucine is added to a casein diet, the growth is drastically depressed. However, addition of isoleucine along with leucine improves the growth pattern but addition of valine, a structurally similar amino acid, along with isoleucine and leucine restores the growth pattern to initial levels (i.e., casein alone). In other words, the antagonism caused by the addition of leucine is overcome by the addition of structurally similar amino acids—isoleucine and valine.

Amino acid toxicity   Two of the essential amino acids namely methionine and tyrosine have been found to produce toxic symptoms and depression in growth of rats when added in excessive amounts to a diet containing casein.

NITROGEN BALANCE

It involves a comparison of the intake of nitrogen in food with the loss of nitrogen from the body in the urine, in faeces and from the surface of the skin. Urinary losses include nitrogen from the breakdown of body tissues (endogenous nitrogen) plus nitrogen from the deamination of absorbed dietary protein in excess of what is needed to build and repair body tissue (exogenous nitrogen).

Nitrogen Equilibrium

When nitrogen intake equals nitrogen loss, a person is in nitrogen equilibrium, indicating that the intake of protein is sufficient to replace any lost tissue but that no growth is taking place.

Positive nitrogen balance   When nitrogen intake exceeds nitrogen losses, an individual is in positive nitrogen balance, indicating that growth is occurring. This positive balance should occur throughout infancy, childhood, adolescence and during pregnancy. It will also occur during recovery from an illness in which protein has been lost.

Negative nitrogen balance   This indicates that nitrogen losses are greater than nitrogen intake. This occurs when the tissues of the body are being broken down at a faster rate than they are being replaced. Prolonged negative nitrogen balance will show up as loss of body weight.

ROLE OF LIVER IN AMINO ACID BALANCE

The liver monitors the absorbed amino acids and adjusts the rate of their metabolism according to bodily needs. When the dietary intake of an essential amino acid is progressively increased, induction of liver degradative enzyme activity usually occurs when intake exceeds requirement which means that the liver destroys the excess of essential amino acids. In the case of non-essential amino acids, the levels of degradative enzymes (e.g. glutamic amino transferase) responsible for their metabolism increase progressively with rising intake.

There is also a conservation of essential amino acids. For example, in a study, varying amounts of lysine were fed to young growing rats. At a daily intake of 100 mg of lysine, gain in body weight was maximal, whereas beyond this level, growth was not stimulated further and the excess lysine was destroyed by the liver.

REGULATION OF BLOOD AMINO ACID LEVELS

Although the liver monitors the passage of some amino acids into the peripheral circulation, amino acids may still be presented to peripheral tissues in amounts greater than that used for protein synthesis.

The plasma levels of amino acids are also affected by dietary carbohydrate intake through a mechanism involving insulin secretion. Shortly after an individual consumes carbohydrate, the concentrations of most plasma amino acids decrease because of deposition in muscle through insulin-mediated transport. The effect is maximal for serum levels of branched-chain amino acids, which can fall as much as 40% after a dose of glucose, whereas some amino acids (e.g. tryptophan) are affected only minimally. The alterations in plasma-free amino acid patterns caused by protein and carbohydrate components of a meal have significance for the availability of amino acids to the peripheral tissues. In particular, the free tryptophan content of rat brain can be elevated by tryptophan administration and in consequence this increases serotonin content of brain. Entry of tryptophan into the cells of the brain is also determined by the plasma levels of other competing neutral amino acids, notably the branched chain amino acids. Consequently, after a meal of carbohydrate, the extensive reduction in plasma levels of branched-chain amino acids results in greater passage of tryptophan into the brain, and more serotonin is synthesized, thus promoting sleepiness.

ROLE OF SKELETAL MUSCLE
IN PROTEIN METABOLISM

Skeletal muscle is the largest tissue in the body. Consequently metabolism of amino acids in this tissue contributes significantly to overall protein metabolism. Muscle is also the main site of metabolism of the branched-chain amino acids (leucine, isoleucine and valine).

Alanine is formed by the transamination between pyruvate derived from glucose and amino groups transferred from amino acids present in muscle. Thus alanine becomes a major carrier of nitrogen from muscle of liver, where its carbon skeleton enters the gluconeogenic pathway, while its amino group is converted to urea or recycled via transamination. Glutamine is the other major carrier of nitrogen from the muscle, and is formed by transamination of glutamic acid in muscle and passes to intestine where about half the nitrogen undergoes transamination to alanine, which is carried to the liver by the portal vein for gluconeogenesis. Following hepatic gluconeogenesis some of the carbon originating in muscle as alanine and glutamine returns to muscle as glucose; this overall exchange between liver and muscle has been named the glucose–alanine cycle. From muscle, a considerable amount of carbon is thus available for metabolism to glucose and thus energy during fasting or other emergencies.

INTEGRATION OF BODY PROTEIN METABOLISM

Integration of body protein metabolism can be summarized as shown in Figure 4.1.

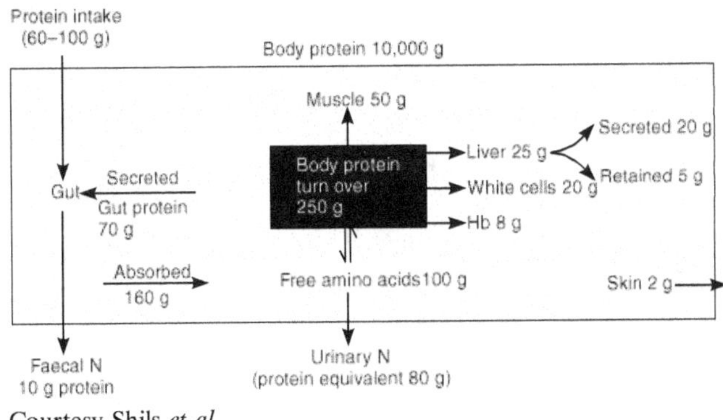

Courtesy Shils *et al.*

Figure 4.1   Summary of protein metabolism

RDA of proteins for different age groups are given in Table 4.1.

Table 4.1   ICMR-recommended dietary allowance for protein

Group	Protein g/day
Man	60
Women	50
Pregnant	50 + 15
Lactation	
0–6 months	50 + 25
6–12 months	50 + 18
Infancy	
0–6 months	2.05/kg
6–12 months	1.65/kg
Children	
1–3 years	22
4–6 years	30
7–9 years	41
Boys	
10–12 years	54
13–15 years	70
16–18 years	78
Girls	
10–12 years	57
13–15 years	65
16–18 years	63

The sources of protein include milk, egg, chicken, meat, cheese, soyabean, nuts and pulses, etc.

PROTEIN QUALITY

The quality of a protein is determined by the kind and proportion of amino acids it contains, i.e., the relation of the amino acids in its molecule to those required for building new tissues. If the protein of the diet is seriously deficient in one or more of the essential amino acids, N equilibrium cannot be sustained, no matter how complete and excellent the diet may be in all other respects. If however, another protein containing the missing amino acid in adequate amounts is added to the diet, N equilibrium and normal nutrition can be established. This capacity of proteins to make good one another's deficiencies is known as their supplementary value.

In 1915 Mendel divided proteins into two classes—those that when fed to rats "allowed growth" and those with which there was "failure of growth".

A knowledge of the amino acid content of proteins can only help to interpret nutritional differences among proteins in terms of their amino acid make-up. Quantitative data regarding the relative digestibility coefficient and nutritive value of proteins, i.e., suitability to meet the protein requirements of the body, can be obtained only through experiments on animals or human beings.

A large number of methods have been proposed by various workers for the evaluation of protein quality. Of these, the most suitable methods for the evaluation of quality of dietary protein are listed in the following section.

Protein Efficiency Ratio

This method was developed by Osborne, Mendel and Ferry in 1919 and is based on the growth of young rats. The diet usually consists of 10% of the protein to be tested and is complete in all other dietary essentials. Groups of albino rats (21-day-old weanling rats) are fed for a period of 4 weeks. Records of the gain in body weight and protein intake of the rats are maintained. The protein efficiency ratio (PER) is calculated using the formula

$$PER = \frac{\text{Gain in body weight (g)}}{\text{Protein intake (g)}}$$

$$= \text{Gain in weight/g protein consumed}$$

Net Protein Ratio (NPR)

This method was introduced by Bender and Doell (1957) and is a modification of the PER method. In this method, an allowance is made for the protein requirements for maintenance. The method consists in feeding a group of weanling rats on a diet containing 10% of the test protein and another comparable control group on a non-protein diet for a period of 10 days.

$$NPR = \frac{\begin{array}{c}\text{Gain in weight (g)} \;+\; \text{loss in weight (g)}\\ \text{of test group} \qquad \text{of non-protein group}\end{array}}{\text{Protein intake (g)}}$$

Digestibility Coefficient (DC)

Dietary proteins are hydrolysed to amino acids during digestion. Proteins differ in their digestibility. In the same period, say 1 to 2 hours, certain proteins like milk and egg proteins are easily digested and converted into amino acids while certain other proteins, like proteins of pulses are only slowly digested to amino acids. Undigested proteins are wasted in the faeces.

The term digestibility coefficient of proteins refers to the percentage of the ingested protein absorbed into the bloodstream after the process of digestion is complete.

$$DC = 100 \times \frac{N_2 \text{ intake} - (N_2 \text{ in faeces} - \text{endogenous faecal } N_2)}{N_2 \text{ intake}}$$

$$= 100 \times \frac{I_n - (F_n - F_e)}{I_n}$$

where, $F_n - F_e$ is the food N_2 lost in digestion.

The DC of proteins is influenced by 1) the presence of indigestible carbohydrates like cellulose and hemicellulose and 2) the presence of proteolytic enzyme inhibitors.

Biological Value (BV)

It was developed by Mitchell in 1925. It measures the quantity of dietary proteins utilized by the animal for meeting its protein needs for maintenance and growth. Groups of albino rats (28-days old) are fed successively on the following diets for a period of 10 days. 1) protein-free diet and 2) 10% protein (to be tested) diet. Urine and faeces are collected by keeping the rats in metabolism cages. BV is the % of retained protein in the body.

$$BV = \frac{(N_2 \text{ digested} - N_2 \text{ lost is metb.})}{N_2 \text{ digested}} \times 100$$

$$N_2 \text{ digested} = N_2 \text{ intake} (I_n) -$$
$$N_2 \text{ in faeces on protein diet} +$$
$$N_2 \text{ in faeces on protein-free diet}$$
$$= I_n - (F_n - F_e)$$

$$N_2 \text{ lost in metabolism} = N_2 \text{ in urine} (U_n) \text{ on protein diet} -$$
$$N_2 \text{ in urine} (U_e) \text{ on protein-free diet}$$
$$= U_n - U_e$$

Therefore,

$$BV = \frac{I_n - (F_n - F_e) - (U_n - U_e)}{I_n - (F_n - F_e)} \times 100$$

If BV is 70%, it means that more than 70% of protein is retained by the body.

Net Protein Utilization (NPU)

Mitchell (1922) introduced NPU. The BV makes no allowance for losses of N_2 in digestion.

$$NPU = \frac{DC \times BV}{100}$$

Miller and Bender (1955) developed a direct method of estimating NPU. Groups of albino rats (28-days old) are used. One group is fed on a non-protein diet while the other groups are fed on the test diet containing test protein at 10% level for 10 days. The food intake is measured. The animals are killed at the end of 10 days and body N_2 is determined.

$$\text{NPU} = \frac{\begin{array}{c}\text{Body } N_2 \text{ of test group} - \text{Body } N_2 \text{ of non-protein} \\ \text{group} + N_2 \text{ consumed by non-protein group}\end{array}}{N_2 \text{ consumed by test group}} \times 100$$

NPU helps in the calculation of the net available protein of the diet. NPU is normally measured with the protein intake at or below maintenance levels. Values determined under other conditions have been termed "operation" (NPUop).

Net Dietary Protein Value (NDPV)

This was proposed by Platt and Muller in 1957.

$$\text{NDPV} = \text{Intake of N} \times 6.25 \times \text{NPUop}$$

Net Dietary Protein Energy Ratio

This was proposed in 1961. The protein content of a food is expressed in terms of the percentage of the energy content provided by protein. The net dietary protein energy ratio is defined as

$$\frac{\text{Protein energy}}{\text{Net dietary intake}} \times \text{NPUop}$$

Standard BV and NPU measurements, which consider only one intake level and zero, tend to overestimate the nutritional quality of some proteins. The best biological estimates of protein quality are provided by the slope of the intake-response line from several points in the range of intakes where the line in linear. If carcass N_2 retentions in animals are used in this way, the index is the relative nutritive value and the line of the test protein is related to a standard (egg or lactalbumin).

Chemical Score

Egg protein contains all essential amino acids in adequate amounts and hence possess the highest nutritive value among dietary proteins. Block and Mitchell (1946) assigned a chemical score of 100 to egg proteins.

The chemical score is the ratio between the content of the most limiting amino acid in the test protein to the content of the same amino acid in egg protein expressed as a percentage. For example, the most limiting amino acid in milk is S-containing amino acid. The chemical score of milk protein $\frac{3.4}{5.5} \times 100 = 65$. The chemical scores of gelatin and zein are zero.

PER, BV, NPU values and the chemical score of some proteins are given in Table 4.2.

Table 4.2   Value of some proteins

Food	PER	BV	NPU	Chemical score
Animal foods				
Egg	4.5	96	91	100
Cow's milk	3.0	84	75	65
Liver	2.9	77	65	66
Meat	2.8	80	76	70
Fish	3.0	85	72	60
Cereals				
Rice	2.0	64	57	60
Wheat	1.7	58	47	42
Pulses				
Bengal gram	1.7	58	47	44
Peas (dried)	1.6	56	45	42

(Contd.)

Table 4.2   (Continued)

Food	PER	BV	NPU	Chemical score
Nuts and oilseeds				
Groundnut	1.7	54	45	44
Soybean	2.0	64	54	57
Coconut	2.5	67	56	52
Sesame	1.7	60	51	40
Incomplete proteins				
Gelatin	0	-	-	0
Zein	0	-	-	0

PDCAA

WHO and the US FDA adopted Protein Digestibility Corrected Amino Acid (PDCAA) score as the official assay for evaluating protein quality. This represents amino acid score after correcting for digestibility.

Proteins that after correcting for digestibility, provide amino acids equal to or in excess of requirements receive a PDCAA of 1.0. Soya protein, egg white and casein (milk protein) have PDCAA of 1.0. Pea protein, rice and wheat have a PDCAA of 0.73, 0.47 and 0.4 respectively.

MUTUAL SUPPLEMENTATION OF PROTEIN

If a protein is deficient in one or more essential amino acids, the nitrogen equilibrium of the individual cannot be maintained. However, if another protein rich in these amino acids is added the quality of the protein improves. This capacity of protein to rectify another's deficiencies is known as supplementary value of proteins.

In any protein the amino acid which is furthest below the standard is known as the limiting amino acid.

When the ingredients or nutrients in food item supplement the ingredients or nutrients in another food item, it is called **mutual supplementation**. For protein synthesis, it has been proved that all essential components have to be present simultaneously. Amino acids that are not used for structural growth are irreversibly metabolized and not available for synthesis.

Cereals, legumes, roots, tubers, leaves, fruits and nuts contain proteins which by themselves are mostly of low biological value. Some of these taken together or supplemented with animal protein may supply a satisfactory amino acid mixture due to mutual supplementation. Effective supplementation occurs only when deficient and supplementary proteins are fed simultaneously or within a short interval of time. The timing in supplementation represents the time factor in protein synthesis.

Table 4.3 lists food limiting in certain amino acids. Grains are limiting in lysine and threonine while legumes and pulses are limiting in methionine and tryptophan. Hence, together they form a good quality protein, able to sustain nitrogen equilibrium.

Table 4.3   Limiting amino acids in vegetable protein foods

Category	Limiting amino acid(s)
Most grain products	Lysine, threonine
Most legumes or pulses	Methionine, tryptophan
Nuts and oilseeds	Lysine
Green leafy vegetables	Methionine
Leaves and grasses	Methionine

Proteins with lysine and threonine have been found to cause marked improvement in PER, e.g. "Modern bread".

The PER of rice proteins improves when fortified with both lysine and threonine.

♔ REVIEW QUESTIONS

1. List the functions of proteins.

2. How are proteins classified?

3. Give the classification of proteins.

4. What are essential amino acids?

5. How can the quality of proteins be determined?

6. Give the RDA for proteins.

7. What are the factors affecting protein requirement?

8. What is meant by limiting amino acid?

CRITICAL THINKING QUESTION ♔

1. In what way are animal protein sources different from vegetable protein sources?

5

ENERGY

Did You Know?

- Carbohydrates can be used for energy and fat storage.

- Eating protein also leads to fat formation.

- The brain uses energy in the form of protein.

- Fasting increases blood ketone levels.

- Ketones are formed primarily from glycerol.

- Mitochondria supply energy within the cell.

- Acetyl-CoA has a central role in energy metabolism.

Energy is that force or power that enables our body to carry on its life-sustaining activities. Metabolism is the total of all those chemical processes in the body by which substances initially in food are changed to other substances. It is these constant, multiple chemical changes in the form of physiological constituents in our bodies that produce energy we live by.

FORMS OF ENERGY

The biologist is interested in energy in five forms.

1. solar

2. chemical

3. mechanical

4. thermal

5. electrical

In plants and animals the various forms of energy are quantitatively interchangeable. Plants use solar energy and convert it to chemical energy by a process known as photosynthesis which is stored by the plant.

Animals get their energy from their food in a chemical form which is derived directly or indirectly from plants. This energy is found in molecules of carbohydrate, fat, protein and alcohol.

Energy taken in as food is used

1. to perform mechanical work,

2. to maintain the tissues of the body and

3. for growth.

The energy in the food is ultimately converted into heat and its dissipation maintains the temperature of the body. The extra heat developed when hard mechanical work is done represents waste that must be eliminated if the temperature of the body is to be kept normal. This is effected by means of sweating.

The unit of energy is the joule (J) and is the energy expended when one kilogram is moved one metre by a force of one Newton. When large amount of energy is concerned, the convenient units are the kilojoule ($=10^3$ J) and the megajoule ($=10^6$ J).

Another measure of energy is the kilocalorie (kcal). This is the amount of heat required to raise 1 kg of water through 1°C. The conversion factors for changing kilocalorie to kilojoules (kJ) is 4.184.

i.e., $$1 \text{ kcal} = 4.184 \text{ kJ}$$

or

$$1000 \text{ kcal} = 4.184 \times 10^3 \text{ kJ}$$

or

$$1 \text{ MJ} = 239 \text{ kcal}$$

$$1 \text{ MJ} = 1000 \text{ kJ}$$

Rates of work or energy expenditure are conveniently expressed in watts (w = J/s). Twelve people sitting and talking in a room produce heat at the rate of about 60 kJ/min. and this is equivalent to a one kilowatt electric fire.

ENERGY VALUE OF FOODS

The energy value of foods is usually determined using the instrument called bomb calorimeter. It consists of a heavy steel bomb with a platinum or gold-plated copper lining and a cover held tightly in place by means of a strong screw collar. A weighed amount of sample, usually pressed into pellet form, is placed in a capsule within the bomb which is then closed except for the oxygen valve charged with oxygen to a pressure of about 300 pounds to the square inch. The oxygen valve is then closed and the bomb immersed in a weighed amount of water. The water is constantly stirred and its temperature taken at intervals of one minute by means of a differential thermometer, capable of being read to one thousandth of a degree. After the temperature of the water is determined, the sample is ignited by means of an electric fuse and on account of the large amount of oxygen present, it undergoes rapid and complete combustion. The heat liberated is absorbed by the water in which the bomb is immersed and the resulting rise in temperature is accurately determined. The thermometer readings are also continued through an "after period" in order that the "radiation correction" may be calculated and the observed rise of temperature corrected accordingly. This corrected rise multiplied by the total heat capacity of the apparatus and the water in which it is immersed gives the total heat liberated in the bomb.

e.g. Wt. of wheat taken = 2 g

Wt. of water in outside vessel = 3,000 g

Wt. equivalent of the calorimeter = 500 g

Initial temperature of water = 24°C

Final temperature of water = 26°C

Rise in temperature of water = 2°C

Heat gained by water and calorimeter   = 3500 × 2

$$= 7,000 \text{ cal}$$

$$= 7 \text{ kcal}$$

2 g of wheat produces 7 kcal

1 g of wheat produces = 3.5 kcal

Calorific value of 1 g wheat = 3.5 kcal

The energy value of a sample of food determined in a bomb calorimeter is known as the heat of combustion, that is, the maximum amount of energy that the sample is capable of yielding when it is completely burned or oxidized. When samples of carbohydrate, fat and protein are burned, the amount of heat produced is always the same for each of these nutrients.

The average calorific values of pure carbohydrates, fats and proteins determined with the bomb calorimeter are:

1 g CHO	4.1 cals
1 g fat	9.45 cals
1 g protein	5.65 cals

An alternate method of measuring food energy is by computing the approximate nutrient composition of a given food. These values are based on the average kilocalorie value of each of the three major energy nutrients.

However, the net heat of combustion in the human is slightly different from that in the bomb calorimeter.

Loss in Digestion

Once the food has been taken, 100% efficiency is not obtained in the digestion or absorption of nutrients. The extent of digestion varies from one nutrient to another and is further influenced by the food in which the nutrient is found.

To calculate the potential energy from carbohydrates, fats and proteins, representative coefficients of digestibility are used to express the percentage of the nutrient that is available. For carbohydrate, fat and protein, the coefficients of digestibility are 0.98, 0.95 and 0.92 respectively showing that 98%, 95% and 92% of each is available for use by the body cells.

Loss of Energy in Metabolism
Due to Incomplete Oxidation

There is no loss in metabolism in the case of carbohydrates and fats. But in the case of proteins, a part of the energy is lost as urea due to incomplete oxidation. This loss has been estimated to be 1.2 cal/g of protein oxidized.

Physiological Fuel Value

In the bomb calorimeter, carbohydrates (CHO) and fats are completely oxidized to CO_2 and water. Protein is oxidized to CO_2, water and nitrogen. Another important error in the use of bomb calorimeter for determining the calorific value of foods of vegetable origin is that the fibre present in foods is burnt and yield energy, while it is not utilized by human beings and a certain percentage is lost in digestion and the nitrogen in proteins is excreted mainly as urea which contains some energy value (Table 5.1). Therefore the physiological fuel values of carbohydrates, protein and fat are 4, 4 and 9 kcal respectively.

Table 5.1   Physiological fuel value

Nutrient	Gross energy value kcal/g	Loss of food energy in digestion %	Energy available after digestion kcal/g	Loss of food energy in metabolism kcal	Physiological energy value of foods kcal/g
CHO	4.1	2	4.0	Nil	4.0
Fats	9.45	5	9.0	Nil	9.0
Proteins	5.65	8	5.2	1.2	4.0

Benedicts' Oxy-calorimeter

Another apparatus used for the determination of the energy value of foods is the oxy-calorimeter devised by Benedict and co-workers. This measures the volume of oxygen required to burn a known weight of the food.

The apparatus consists of a combustion chamber in which the weighed sample is burnt, a soda lime container for absorption of carbon dioxide, a spirometer for measuring the oxygen used and a motor blower unit for circulating gas mixture using this instrument, the amount of oxygen consumed in burning 1 g of pure CHO, fat or protein can be determined.

Relation between oxygen required and calorific value

1 g of CHO requires 0.8 of O_2 for complete oxidation and yields 4.1 k cal

1 g of fat requires 2.2 of O_2 for complete oxidation and yields 9.5 k cal

1 g of protein requires 1.2 of O_2 for complete oxidation and yields 5.5 k cal

Therefore,

1 litre of O_2 oxidizes 1.25 g of carbohydrates and gives 5 kcal heat

1 litre of O_2 oxidizes 0.49 g of fat and gives 4.5 kcal heat

1 litre of O_2 oxidizes 0.83 g of protein and gives 4.6 kcal heat

One litre of oxygen oxidizing carbohydrates, fat or protein produces nearly the same amount of heat, i.e., 4.5 to 5 kcal. This is the basis for indirect determination of energy value of foods from oxygen consumed.

BASAL METABOLISM

The energy metabolism of a subject at complete physical and mental rest having normal body temperature and in the post-absorption state (i.e., 12 hours after the intake of last meal) is known as basal metabolism. It is the sum of all internal chemical activities that maintain the body at rest. It is a measure of the energy required by these activities of resting tissue. Certain small but vitally active tissues—brain, liver, gastrointestinal tract, heart, kidney—together make up less than 5% of the total body weight, yet they contribute to about 60% of the total basal metabolic processes.

Measurement of Basal Metabolism

Basal metabolism can be measured in the same way that we measure the energy in food by direct calorimetry. More common however, is indirect calorimetry which measures oxygen used rather than heat produced to assess energy used.

Direct calorimetry To measure BMR involves the measurement of the heat given off by the body in a human calorimeter or respiration chamber, a small insulated room that operates on the same principle as the bomb calorimeter. By measuring the change in temperature of a known volume of water circulating in pipes on the top and walls of the chamber, it is possible to determine the amount of heat produced by a subject inside it.

This equipment consists of an air-tight copper chamber insulated by wooden walls with air space in between. A folding bed, chair and table are provided in the chamber. A man can comfortably stay in the calorimeter for a few days. A small opening is provided at the two ends for passing food and drinks and removing excreta. The chamber is ventilated by a current of air, the CO_2 and water given off are removed by soda lime and sulphuric acid respectively. Oxygen utilized by the subject is replaced by introducing known amounts of oxygen through gas meter into the chamber. The amount of O_2 and CO_2 produced can be calculated from the above. The heat produced is measured accurately by circulating a current of water through copper pipes and measuring the amount of circulating water with the increase in temperature of water.

e.g. Adult weighing 65 kg

Amount of heat output in 24 hours = 2400 kcal

Amount of oxygen conserved in 24 hours = 500 l

Heat output per litre of oxygen consumed = 4.8 kcals

The respiratory quotient (RQ) can be measured which is the ratio of CO_2 produced to O_2 consumed. From the RQ we can determine whether CHO, fat, or protein was used for energy. If CHO is the sole source of fuel the RQ is 1. Fat has an RQ of 0.7, and protein RQ of 0.8 depending on the amino acid mixture. Under basal conditions in which both fatty acids and glucose are being used as energy sources, the RQ is usually 0.82.

Indirect method Basal metabolism is usually determined using the apparatus of Benedict and Roth. The apparatus is a closed circuit system in which the subject breathes in oxygen from a metal cylinder of about 6 litre capacity and CO_2 produced is absorbed by soda lime present in the tower. The oxygen cylinder floats on water present in an outer tank. The subject wears a nose clip and breathes through a mouthpiece, the oxygen present in the cylinder for a period of 6 minutes. The volume of O_2 used is recorded on a graph paper attached to a revolving drum by a pen attached to it. Since the subject is in a post-absorptive state, RQ is assumed to be 0.82 and the calorific value of one litre of O_2 consumed is taken as 4.8 kcal.

Example   Subject: Adult male, 50 kg body weight

Oxygen consumed in 6 minutes   = 1,100 ml

Heat produced in 6 minutes   = 1,100 × 4.8

= 5.28 kcal

Heat produced in 24 hours = 5.2 × 10 × 24 = 1,267 kcal

The basal metabolism of the individual for 24 hours

= 1,267 kcal

Standards for Basal Metabolism

Studies carried out by various workers have shown that basal metabolism is most closely related to the body surface area and less directly related to either the weight or height of the individual. The body surface area can be calculated according to the formula of Dubois and Dubois given below.

$$A = W^{0.425} \times H^{0.725} \times 71.84$$

Factors Affecting the Basal Metabolic Rate (BMR)

Body size   The BMR is closely related to the body surface area. It is directly proportional to the body surface area.

Age   The BMR is higher in infants and young children than in adults. The decline in basal energy needs between 25 and 35 amounts to only 35 kcal/day. Those who fail to adjust caloric intake put on weight.

Sex Females have 5% lower BMR than males. Women characteristically develop more adipose tissue and less musculature than men.

Body composition The BMR is directly related to the lean body mass. Persons with well-developed muscle will have a higher BMR than obese persons whose higher body weight is due to adipose tissue. All body tissues are metabolically active, constantly being broken down and repaired and participating in vital functions. Muscle, brain, gland and liver are active metabolically (consume large amounts of O_2) while bone adipose tissues are relatively inactive.

Climate In persons living in tropical climates, the BMR is about 10% lesser than those living in temperate zones.

Specific dynamic action (SDA) Food has a stimulating effect on BMR. The BMR of a person increases by about 8% when food is ingested.

Undernutrition and starvation Prolonged undernutrition or starvation causes a reduction of about 10–20% in BMR. This reflects body's adaptive effort to conserve energy.

Sleep The BMR in sleep is about 5% less due to muscular and emotional relaxation.

Fever Fever increases the BMR. For every 1°F rise in body temperature, the BMR increases by about 7%. Therefore there is an increased need for energy.

Physical activities Physical activity increases the BMR.

Fear and nervous tension These increase BMR by increasing the secretion of adrenaline.

Thyroid Hypothyroidism decreases BMR up to 30% and hyperthyroidism may increase BMR up to 100% depending on the severity of the condition.

TOTAL ENERGY NEEDS OF THE BODY

The energy needs of the body are affected by various factors.

Activity

Energy costs for activity include the amount of energy needed for all the muscles involved in the activity plus a small amount of energy to take care of the increase in heart rate and breathing that takes place during strenuous activity. For many activities, energy costs depend on body size and the relative severity of the exercise.

The actual cost of various activities has been determined by a series of tests that measure the amount of oxygen consumed and therefore the energy used in performing a specific activity.

Influence of Various Body Weights on Energy Requirements for Work

The energy required for walking at a particular speed will depend on the body weight of the individual. Studies have shown that for walking at the rate of 2 miles per hour an individual with body weight of 200 lbs will require twice as much energy as an individual weighing 80 lbs and one-half times as much energy as an individual weighing 120 lbs.

Energy Expenditure in Relation to Intensity of Muscular Work

1. Light work—150–294 kcal/hour.

    Light industry, carpentry, military drill, domestic work, painting, driving, etc.

2. Moderate work—300–444 kcal/hour.

    Agricultural work, route march, tennis, cycling (6 miles/hour)

3. Heavy work—450–594 kcal/hour

    Coal mining, foot ball, hockey

4. Very heavy work—Over 600 kcal/hour

   Lumber work, running, swimming, hill climbing

Thermic Effect of Food or
Specific Dynamic Action of Food

Rubner observed that carbohydrates, fats, and proteins fed to a fasting dog stimulated the energy metabolism over basal level to varying extents. He found that in a fasting dog requiring 400 kcal feeding of 100 g of CHO produces 425 kcal, 44.4 g of fat produces 416 kcal and 100 g of protein produces 520 kcal of heat. Therefore, the thermic effect of food refers to the stimulation in metabolism and therefore the production of heat that occurs from 1 to 3 hours. after a meal as the result of the presence of food in stomach and intestine and nutrients in the bloodstream. The increase in energy cost because of this thermogenesis amounts to about 10% of the total for basal metabolism and activity. Therefore, to estimate total energy needs, it is necessary to add an additional 10%.

The SDA of proteins is the highest (about 30%) while that of carbohydrates and fats is only 6% and 4% respectively. The SDA of a mixed diet containing 62.5 g of carbohydrates, 10 g of fats and 10 g of proteins is about 8%.

According to Krebs, two main factors are responsible for the high SDA of proteins.

1. The energy required for deamination of amino acids which again is derived by the oxidation of other metabolites and

2. The energy required for the synthesis of urea which is obtained by the oxidation of other metabolism.

The phenomenon of thermogenesis ("waste heat") is influenced by many factors. Epinephrine, norepinephrine, nutrient deficiencies, etc. increase heat production. Thermogenesis varies in amount and is dependent on metabolic pathways involved and these in turn on the availability of certain substrates which ultimately depend upon the nutrientsingested.

ENERGY NEEDS OF SPECIAL GROUPS

Pregnant women   The additional energy needs of pregnancy represents the energy in the developing foetus, which is calculated to be approximately 40,000 kcal over the 9 months of pregnancy. It is suggested that the energy intake of the expectant mother be increased by 250 kcal/day to permit the growth of the foetus.

Lactating women   Assuming an average milk production of 750 ml/day, the added energy cost for lactation is 600–700 kcal.

Infants   The energy requirement at birth is approximately 110 kcal/kg of body weight and declines as the baby grows.

Children and adolescents   Includes the need for growth. After age 10, estimated energy needs are based on resting energy expenditure (REE), a factor of 1.7 for girls and a factor of 1.8 for boys to include needs for activity. The needs for growth represent an increase of less than 3% over basal needs plus activity needs.

ESTIMATION OF TOTAL ENERGY NEEDS

The total energy requirement of an individual is made up of three main components basal or resting metabolism, acitivity and thermic effect of foods. All these components vary depending on a variety of factors. To estimate energy needs, it is therefore possible to calculate each of these three components separately, adjust for as many factors that influence need as possible, and then add them together.

Some methods for estimating total energy requirements are

Factorial method = Basal metabolism (A) + activity (B) + thermic effect of food (C).

The method of calculating basal metabolism, and energy cost of activity is explained below.

1. Basal metabolism can be calculated using one of the following methods.

    Method 1

    For males,

    Basal energy = Body weight (kg) × 1.0 × 24 kcal/hour

    For females,

    Basal energy = Body weight (kg) × 0.9 × 24 kcal/hour

    Method 2   FAO/WHO/UN

    For males,

    Basal energy = 11.6 × weight(kg) + 879

    For females,

    Basal energy = 8.7 × weight(kg) + 829

2. Calculate activity costs based on a record of activity over a 24-hour period from Table 5.2.

Table 5.2   Energy cost of exercise

Activity	Prime (hour) [A]	Energy cost kcal/kg/hour [B]	kcal/kg [A] × [B]
Dressing	1.5	0.7	1.05
Sitting	6.0	0.4	2.4
Walking (3 mph)	2.0	2.0	4.0
Standing	1.0	0.5	0.5
Sleeping	8.0	–	–
Eating	0.5	0.4	0.2

Energy cost = Body weight (kg) × 14.85

Add basal energy costs and activity costs

3. Calculate thermic effect of food

Total basal + activity × 10%

Calculation of amounts of proteins, fats and carbohydrates oxidized in the body From the quantity of nitrogen excreted in urine and the quantity of CO_2 produced and oxygen consumed in 24 hours, it is possible to calculate the amount of proteins, lipids and carbohydrates oxidized in the body.

RECOMMENDED ALLOWANCES FOR CALORIES

In keeping with usual practice the committee (ICMR) defined the energy requirement of an adult man and woman in terms of a reference man and reference woman.

Reference man Reference man is between 20–39 years of age and weighs 55 kg. He is free from diseases and is physically fit for active work. On each working day he is employed for 8 hours in occupation that usually involves moderate activity. While not at work, he spends 8 hours in bed, 4–6 hours sitting and moving about and two hours in walking and in active work or in household duties.

Reference woman Reference woman is between 20–39 years of age, is healthy and weights 45 kg. She may be engaged for 8 hours in general household work, in light industry or in other moderately active work. Apart from 8 hours in bed, she spends 4–6 hours sitting or moving around only through light activity and 2 hours in walking or in active recreation or in household duties.

Table 5.3 summarizes the RDA for energy for all age groups.

Table 5.3   ICMR-recommended allowances of energy

Group	Energy kcal
Man	
Sedentary work	2425
Moderate work	2875
Heavy work	3800
Women	
Sedentary work	1875
Moderate work	2225
Heavy work	2925
Pregnancy	+300
Lactation	+550
0–6 months	+400
6–12 months	
Infants	
0–6 months	108/kg
6–12 months	98/kg
Children	
1–3 years	1240
4–6 years	1690
7–9 years	1950
Boys	
10–12 years	2190
13–15 years	2450
16–18 years	2640
Girls	
10–12 years	1970
13–15 years	2060
16–18 years	2060

Sources

Oils, fats, nuts and oilseeds, sugar and other carbohydrate are rich sources of energy. Fruits and vegetables rich in moisture are poor sources of energy.

⚘ REVIEW QUESTIONS

1. Give the units of energy.

2. How can the energy value of foods be determined?

3. Define gross fuel value and physiological fuel value.

4. How is BMR measured?

5. What are the factors affecting BMR?

6. Define reference man and woman.

CRITICAL THINKING QUESTIONS ⚘

1. Can protein lead to fat formation?

2. Can the energy value of foods be determined from the oxygen consumed?

6

PROTEIN ENERGY MALNUTRITION

Did You Know?

- The prevalence rate of severe degree of PEM is 3–5%.

- Maternal malnutrition prior to or during pregnancy causes low birth weight infant.

- Marasmus causes severe growth retardation.

- PEM is more frequent among infants and young children.

- Hepatomegaly is seen in PEM child.

- PEM during pre-school stage can have a permanent damage on the brain.

Protein energy malnutrition (PEM) describes a range of clinical disorders resulting from a lack of varying proportions of energy and protein.

AETIOLOGY

PEM occurs characteristically in children under five years of age, wherever the diet is poor in protein and energy.

The factors which precipitate PEM can be classified as those relating to the level of protein in the diet, the age, pathological status of the child and maternal nutritional level and finally the environment.

HOST FACTORS

Level of Protein

If the quantity of protein is too low or if the biological value of the protein is low or if both quantity and quality of a protein is low, the requirements of protein is not met.

Another main factor influencing protein deficiency is the proportion of calories in the diet.

The biological value of the protein depends on the relative amounts of essential amino acids present and biological availability following digestion.

Age

Age influences the appearance of PEM in several ways. Since nutritional requirements are directly related to the rate of growth, the younger the child the greater the protein requirement per unit of body weight.

Since small children are not able to masticate foods such as meat, unless they are specially prepared, their diet is customarily limited to gruels and other foods which are easy to swallow but poor in nutrition.

Pathological Status of the Child

Infectious processes such as infectious diarrhoea is an important precipitating or aggravating factor. Non-infectious processes such as malabsorption syndrome and metabolic abnormalities of endocrine origin may also cause PEM.

Maternal Malnutrition

Malnutrition of the mother during pregnancy is more likely to produce an underweight newborn baby. This compounded with insufficient food results in PEM.

Environment

Among the environmental factors limiting the availability of protein are

a. Physical and geographical factors influencing food production such as climate and characteristics of soil.

b. Biological factors such as the varieties of plants or animals which can be supported in a given area.

c. Socio-economic factors such as knowledge of good agricultural practices, the provision for land distribution, efficiency of conservation, processing and storage of foods.

d. Ignorance by itself or associated with poverty leading to poor infant and child rearing practices.

e. Overcrowded or unsanitary living conditions leading to frequent infections.

Of the environmental factors affecting food consumption, the socio-economic factors are of greater importance. The purchasing power of the population and their habits, practices, beliefs and taboos regarding food particularly the feeding of young children are often the most important factors affecting the protein requirement of the host.

CLASSIFICATION OF PEM

The classification of PEM as grades is as follows.

- Grade I — 75–90% of the reference weight for age.

- Grade II — 60–74% of reference weight.

- Grade III — less than 60% of reference weight.

Use of the upper arm circumference allows differentiating between a moderate and severe malnourished child and one with a better nutritional condition.

Mild and Moderate PEM

The main clinical feature of mild and moderate PEM is weight loss. A decrease in subcutaneous adipose tissue may become apparent. When PEM is chronic, children show growth retardation in terms of height.

Severe PEM

The diagnosis is principally based on dietary history and clinical features. Marasmus is associated with severe food shortage, prolonged semistarvation, early weaning or infrequent feeding of infants and kwashiorkor with late weaning and poor protein intake. The clinical presentation depends on the type, severity and duration of the dietary deficiencies.

MARASMUS AND KWASHIORKOR

Marasmus is due to a continued restriction of both dietary energy and protein as well as other nutrients. The urban influences which predispose to marasmus are a rapid succession of pregnancies and early and abrupt weaning, followed by dirty and unsound artificial feeding of the infants with very dilute milk products, given in inadequate amounts to avoid expense. Thus the diet is low in both energy and protein. In addition, poor housing and lack of equipment make the preparing of clean food almost impossible.

Repeated infections develop, especially of the GI tract; these the mother often treats by starvation for long periods, the infants receiving water, rice-water or other non-nutritious fluid.

In marasmus, weaning has often been early. The mother may be induced to stop breast feeding for various reasons including the presence of infections in herself or in the infant. Unfortunately she may have been influenced unwisely by advertisements in the press or radio which advocate for commercial reasons the advantage of artificial food products. A frequent reason for stopping breast feeding is the beginning of another pregnancy. Another reason in urban areas is the necessity or desire to return to paid work.

Clinical Features of Nutritional Marasmus

There is a failure to thrive, irritability, fretfulness or alternatively apathy. Diarrhoea is frequent, many infants are hungry, but some are anorexic, the child is wizened and shrunken and there is little or no subcutaneous fat (Figure 6.1). There is often dehydration. The weight is much below the standard for age. The temperature may be subnormal. If the disease is of long duration, the length of the child is also below the standard, but less so than the weight. There is usually watery diarrhoea with

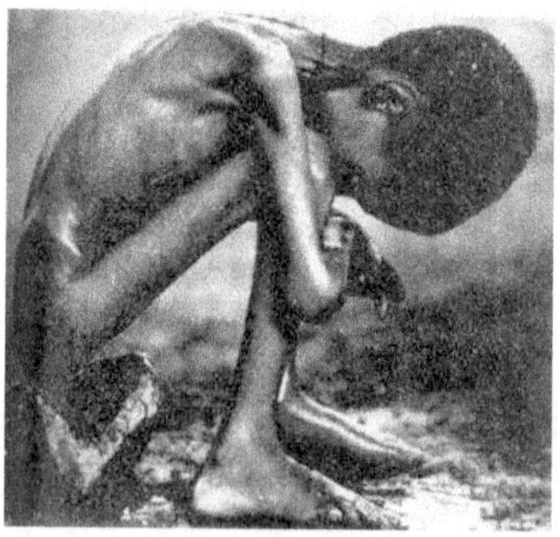

Figure 6.1   Protein energy malnutrition

acid stools. If infective gastro-enteritis is added, the diarrhoea is severe. The abdomen may be shrunken or distended with gas. Because of the thinness of the abdominal wall, peristalsis may be easily visible. The muscles are weak and atrophic and this together with the lack of subcutaneous fat, makes the limbs appear as skin and bone (Figure 6.2a).

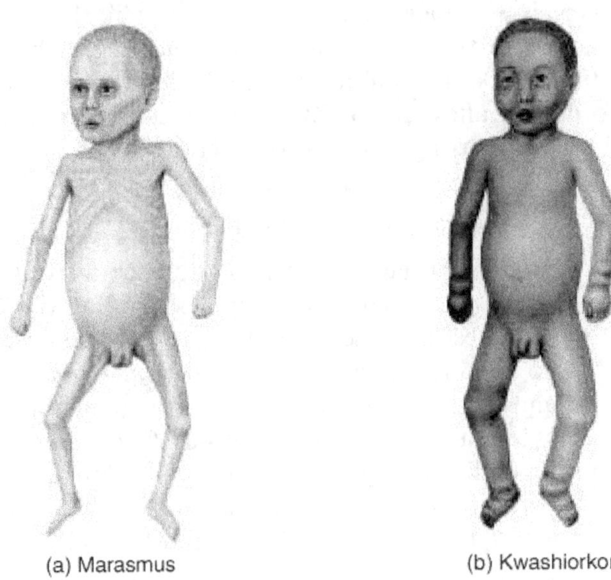

(a) Marasmus   (b) Kwashiorkor

Figure 6.2   Protein-calorie malnutrition

The skin and mucous membrane may be dry and atrophic, but the characteristic changes found in kwashiorkor are not usually present. Evidence of vitamin deficiency may or may not be found. Psychological disturbances resulting from a lack of a mother's love and care can depress the appetite and hence may be a factor in the causation of marasmus.

Kwashiorkor   arises when after a prolonged period on the breast the child is weaned onto the traditional family diet; this may be low in protein because of poverty, insufficient land and poor agricultural practices. There is no supplement of milk or if present, may be a totally inadequate one. Custom, sometimes reinforced by taboos, determines that the limited supply of foods of animal origin is given mainly to the men of the family.

Kwashiorkor is frequently precipitated in epidemic proportions, by outbreaks of febrile illness such as malaria, measles or gastro-enteritis. Both marasmus and kwashiorkor arise as a result of poverty and ignorance. Even if food is available and there is the money to buy it, many mothers have received no satisfactory instruction in infant feeding. Egg, fish, meat and sometimes milk may not be given to children because custom or taboo do not allow them. Kwashiorkor is mainly a disease of rural areas occurring in the second year of life.

Clinical Features of Kwashiorkor

There is oedema together with a failure to thrive, anorexia, diarrhoea, and a generalized unhappiness or apathy. A infection often precipitates the onset and may be the reason for bringing the child to the doctor. Failure of growth is an early sign, though oedema and the presence of some subcutaneous fat make the weight loss less striking than in marasmus. Oedema may be slight or gross depending partly on the amount of salt and water in the diet. It may be distributed over the whole body including the face but is usually more marked on the lower limbs.

Ascitis and pleural effusions are usually slight and, if detected clinically, suggest the presence of an infection.

The characteristic dermatosis consists of areas of desquamation and areas of both hypo- and hyperpigmentation. The skin first becomes thickened and then cracks and peels leaving denuded areas and ulceration. This looks like crazy paving and in very severe cases looks like a burn. Ulceration can occur over pressure points and deep cracks appear in skinfolds. The lower limbs, buttocks and perineum are usually most affected. This may be associated with skin infections and vitamin deficiency symptoms like pellagra.

The hair is sparse, soft and thin. There may be changes in pigmentation with diffuse patches or streaks which may be red, blond or grey in colour. The hair root also changes in structure as there is a reduced diameter of the bulb.

Angular stomatitis, cheilosis and a smooth atrophic tongue are commonly seen as is ulceration around the anal region.

Watery diarrhoea or large semi-solid, acid stools are usual. Hepatomegaly is seen.

The muscles are always wasted and as a result many children regress in their physical development and may no longer be able to walk or crawl (Figure 6.2b).

Apathy is a characteristic feature and the child appears constantly unhappy. Neurological features are unusual, but some children during recovery have unexplained tremors resembling parkinsonism.

Biochemical and Metabolism Disorders

These disorders arise as a direct or indirect result of an insufficient supply of energy and of proteins.

Body composition A high body water content, loss of fat stores and loss of protein from the wasted muscles and other tissues greatly alters the body composition of the child. Children with PEM are underweight with an excessive loss of body protein but not collagen.

General metabolism The metabolism rate is reduced. The cells are able to utilize nutrients to restore lost tissue provided electrolyte disturbances are corrected and infections are treated.

Protein metabolism Plasma concentration of essential amino acids especially branched-chain amino acids and tyrosine are low but some of the non-essential amino acids are higher than normal.

The plasma albumin concentration is low owing to a failure of synthesis in the liver. A low albumin level is in part responsible for the oedema present.

Plasma IgG is often raised if infections are present but other immunoglobulins are normal.

The blood urea is low and may be about 1 mmol/l. This reflects reduced protein intake.

Plasma retinal-binding protein is also low. Urinary creatinine is also reduced, reflecting decreased muscle mass.

Lipid metabolism A fatty liver is characteristic of kwashiorkor but is unusual in marasmus. The excess fat in the liver is triglyceride. In kwashiorkor, plasma triglyceride and plasma cholesterol are low due to a decreased ability of the liver cells to mobilize lipid in the form of lipoproteins.

Carbohydrate metabolism Blood glucose is usually normal. However, hypoglycaemia may occur. Glucose tolerance may be impaired.

Electrolyte and water metabolism A deficiency of potassium arises as a result of diarrhoea. Plasma levels of potassium is low. A deficiency of magnesium can also occur due to increased losses.

Sodium deficiency can occur due to an inability of the kidneys to conserve sodium.

Acidosis can occur and is due to poor circulation and consequent tissue hypoxia. Alkalosis occurs sometimes and is associated with potassium depletion and a failure of the kidneys to excrete bicarbonate.

Changes in the Organs and Organ Systems of the Body

Digestive organs The cells of the pancreas and the intestinal mucosa atrophy, and cannot produce digestive enzymes in normal amounts.

The quantity of enzymes, especially disaccharidases, amylase, trypsin and lipase, are low with impaired activity. Absorption of essential nutrients is also low.

Liver   Fat accumulates in small droplets within liver cells situated at the periphery of the lobules. The droplets increase in size and extend from the periphery to the centre of the lobules. However, liver failure is unusual.

Cardiovascular system   Atrophy of the heart muscle is seen in severe chronic PEM. This leads to a reduced cardiac output and a poor circulation. In severe cases, the extremities are cold and cyanosed and the pulse is small. The ECG shows low-voltage changes in the QRS complex.

Kidneys   The glomerular filtration rate is low due to dehydration or reduced cardiac output. Mild albuminuria may be seen and the concentrating power of the kidney is poor.

Immune system   The lymphoreticular organs are atrophied along with a delayed or absent tuberculin response, reduced number of lymphocytes (T cells), etc. Bacterial action of neutrophils is also impaired. There is a depression of cell-mediated immunity.

Prevention and Treatment

Preventive measures for kwashiorkor and marasmus include nutritional supplementation, treating secondary infection and other life-threatening situations.

Nutritional Requirements

The rationale behind dietary management to treat PEM is to provide levels of protein and energy which will meet nutritional requirement as well as replenish lost growth.

Energy   The child should be given 150 to 200 kcal/kg body weight/day for the existing weight. It is important that adequate calories be provided in the diet.

Protein   5 g of protein per kg body weight per day should be given. The protein provided should be of good quality, be able to regenerate serum albumin and replace lost tissue.

Fats   40% of total calories can be from fat. However care should be taken, as unsaturated fats can worsen diarrhoea.

Electrolytes   Potassium chloride and magnesium chloride should be added.

Vitamins   All vitamins are to be added in required amounts. If vitamin A deficiency is present, oral dose of 50,000 IU of vitamin A should be given.

Treatment of Infection, Dehydration, Oedema and Hypothermia

Diarrhoea, chest infection and skin infection should be treated with appropriate antibiotics.

Children with dehydration in marasmus should be administered fluids with oral rehydration solution either nasogastrically or by intravenous fluid therapy in very severe cases.

Children with frank oedema in kwashiorkor should be corrected with a proper protein diet and maintenance of electrolyte balance.

Children with marasmus have a low body temperature and the child should be covered well and made to sleep with the mother.

Dietary Management

The child should be given sufficient calories and proteins in gradually increasing amounts without provoking digestive upsets. Easily digested foods should be given.

The symptoms including mental apathy should disappear within a week.

PREVENTIVE MEASURE FOR PEM

The following steps will aid in health promotion and prevention of PEM.

- Increasing the nutrient intake and improving nutritional status of the mother during pregnancy and lactation.

- Promoting breast feeding

- Nutrition education

- Early diagnosis and treatment

- Complete implemention of feeding programmes like the ICDS (Integrated Child Development Services).

REVIEW QUESTIONS

1. What are the clinical and biochemical features of marasmas and kwashiorkor?

2. Explain the aetiological factors of PEM.

3. Describe the treatment of kwashiorkor.

CRITICAL THINKING QUESTIONS

1. How is age related to the appearance of PEM?

2. Can environmental factors limit the protein availability?

7

FAT-SOLUBLE VITAMINS

Did You Know?

- People who use mineral oil as a laxative at mealtimes are susceptible to fat-soluble vitamin deficiencies. Mineral oil is not absorbed by the intestine. During its passage, mineral oil can dissolve fat-soluble vitamins and pull them into the colon for eventual elimination.

- It is extremely unlikely that all vitamins have been discovered. There is a slight chance that one more vitamin remains to be discovered, but for the most part nutrition scientists feel they have isolated all the vitamins necessary for human health.

- Vitamin D improves calcium absorption. When in the active from, vitamin D, through a variety of mechanisms, can improve calcium absorption from the small intestine.

- Vitamin A is important for night vision. Vitamin A enhances night vision by participating in the formation of the compound rhodopsin.

- Vitamin A toxicity may appear at five times the RDA. It has a high potential for causing toxicity, especially in pregnant women and children.

- Vitamin D is a hormone, synthesized by sunlight. Aided by the action of sunlight, the skin produces vitamin D, which can then be metabolized by the liver and kidneys to form the active vitamin D hormone.

- Vitamin K is important for blood clotting. Vitamin K is able to add hydroxyl groups (–OH) to one type of amino acid, in turn imparting calcium-binding properties to the proteins that contain those acids.

- Antibiotics can reduce the growth of bacteria in the small intestine and colon that synthesize a form of vitamin K that we absorb and use. Thus antibiotic use provokes a vitamin K deficiency.

- Vegetables are good sources of vitamin K. This is another reason why we should "eat our vegetables". A safe dosage for a supplement for most vitamins is generally 50% to 150% of the RDA.

VITAMIN A

Night blindness was a well-recognized disease in ancient Egypt. The cure as expressed in the Papyrus was to apply topically to the eyes juice squeezed from cooked liver. The ancient Greeks recommended the ingestion of cooked liver as well as its topical application.

In 1912, Osborne and Mendel reported that animals grew normally on diets containing milk fat but failed to grow when milk fat was withdrawn. Others observed that this growth failure was followed by eye disease and then cured by feeding butter, cod liver oil or egg yolk. In 1913, E.V. McCollum and Davis contained this factor and called it "fat-soluble A". Moore in England showed that β-carotene was converted biologically to a colourless form of vitamin A, which was then stored in the liver. In 1930, Karrer and his colleagues determined the structures of both vitamin A and β-carotene. Five years later Wald identified the chromophore of visual pigments as retinene.

CHEMISTRY

Vitamin A is a pale yellow substance that is soluble in fat or fat solvents. Vitamin A and the carotenoids consist of four isoprenoid units joined in a head-to-tail manner and contain five conjugated double bonds.

They are soluble in most organic solvents but not in water. Vitamin A and the carotenoids are sensitive to oxidation, isomerization, and polymerization when dissolved in a dilute solution under light in the presence of oxygen.

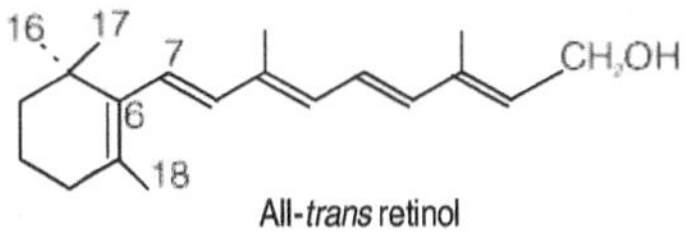

All-*trans* retinol

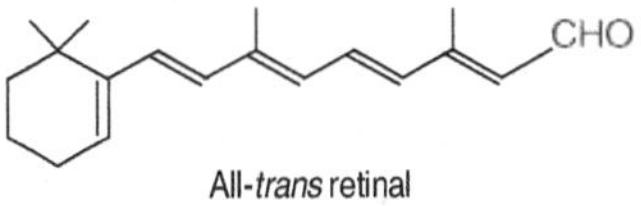

All-*trans* retinal

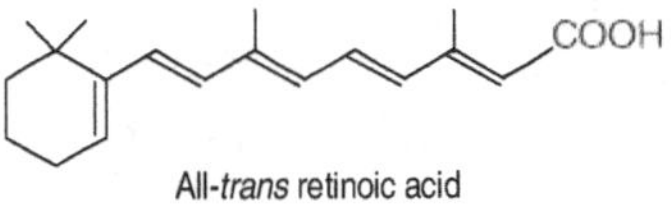

All-*trans* retinoic acid

BIOLOGICAL ACTIVITY

The primary unit of biological activity for vitamin A is 1 µg of all-*trans* retinol whether present as the free alcohol or as one of several natural or synthetic fatty acyl esters.

$$1 \text{ Retinol Equivalent (RE)} = 1 \text{ µg retinol}$$
$$= 6 \text{ µg } \beta\text{-carotene}$$
$$= 12 \text{ µg other provitamin carotenoids}$$
$$= 3.3 \text{ IU retinol}$$
$$= 9.9 \text{ IU } \beta\text{-carotene}$$

FUNCTIONS

Cellular Differentiation

In vitamin A deficiency, mucus-secreting cells are replaced by keratin-producing cells in many tissues of the body. It has been demonstrated that cell nuclei contain four receptors for retinoic acid termed RAR α to γ (retinoic acid receptor). Retinoic acid is transported to the nucleus on cellular retinoic-acid-binding protein (CRABP) where it interacts with one more of the RARs. The activated RAR interacts with the different hormone response elements of appropriate genes to influence transcription.

Vision

In the outer segment of rod cells in the retina, 11-*cis* retinal combines with a specific lysine residue of the membrane-bound protein opsin, to yield rhodopsin. Similar complexes exist in the cone cells to give three specific iodopsins that absorb light at 420 nm (blue cones) 534 nm (green cones) and 563 nm (red cones). When a photon of light strikes the dark-adapted retina, the 11-*cis* bond of retinal in rhodopsin is isomerized to the all-*trans* form. This isomerization destabilizes rhodopsin which passes through a series of different conformational states and ultimately dissociates into all-*trans* retinal and opsin.

All-*trans* retinal may be isomerized back to 11-*cis* form in the rod's outer segment by light in the presence of certain phospholipids. In the dark, however, 11-*cis* retinol is formed by the action of all-*trans* 11-*cis* retinyl ester isomerase, a membrane-bound enzyme in retinal pigment, epithelial cells.

The light-activated transformation of rhodopsin ultimately results in a reduction in the sodium ion current into the rods' outer segment, which induces hyperpolarization of the membrane.

This cascade is explained in Figure 7.1. Light converts rhodopsin to the active metarhodopsin II. The latter induces the conversion of guanosine diphosphate to guanosine triphosphate carried on a protein called transducin. Transducin–GTP complex activates phosphodiesterase which in turn hydrolyses cyclic guanosine monophosphate to guanosine monophosphate. As the concentration of cGMP falls, the sodium channel closes, leading to membrane hyperpolarization. The cascade is turned off by the decay of metarhodopsin II to opsin.

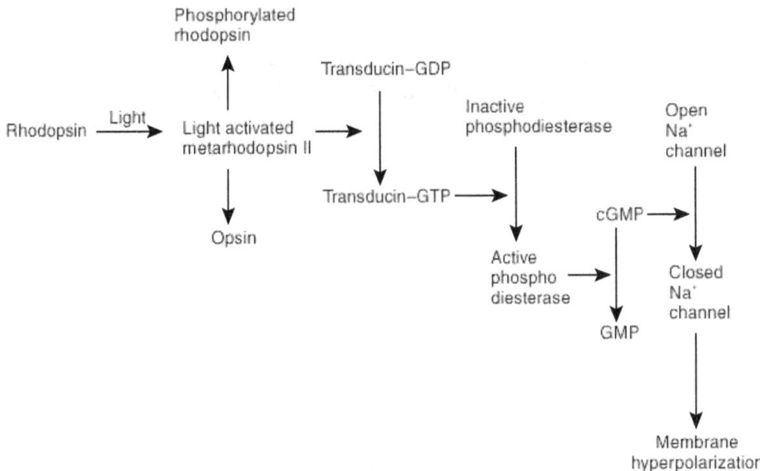

Figure 7.1   Rhodopsin cascade

Immune Response

Vitamin A was termed as the "anti-infective" vitamin. During deficiency, both specific and non-specific protective mechanisms are impaired. Humoral response to bacteria, cell-mediated immunity, mucosal immunity, natural killer cell activity and phagocytosis are all impaired.

The activity of T lymphocytes and in particular of T-helper cells seems to be mainly affected by vitamin A depletion.

The synthesis of goblet cell mucins is reduced by vitamin A deficiency.

Growth

Growth failure occurs during vitamin A deficiency.

Epithelial Cells

When vitamin A is absent, the cells become keratinized and lack cilia and lose their natural property.

Bone Development

Vitamin A is essential for normal bone growth. During deficiency, bones fail to lengthen and are thick but weak and the remodelling process is poorly controlled.

Reproduction

Vitamin A is necessary for normal reproduction in rats. In the absence of vitamin A, the male rat fails to produce sperm cells and the female absorbs the foetus back into the body.

Other Functions

There is a loss of appetite in vitamin A, deficiency and is caused by changes in the taste buds. Wound healing is also impaired as a result of changes in the epithelial cells.

Vitamin A is also necessary for the synthesis and secretion of various cytokines and growth factors.

METABOLISM

Digestion and Absorption

Most of the preformed vitamin A in food combines with the fatty acid, palmitic acid, to form retinyl palmitate. Retinyl palmitate is split by enzymes from either the pancreatic juice or mucosal cells to form free retinol. Bile is important for the uptake of retinol and is essential for carotenes. Once retinol is within the mucosal cell, it combines with a fatty acid (usually palmitic acid) and is incorporated into the small transport particles of fat, i.e., chylomicrons. This, through the lymphatic circulation, eventually reaches the liver.

Because vitamin A is fat-soluble, factors that promote the absorption of fat enhance vitamin A absorption, and those that depress fat absorption depress vitamin A absorption. Vitamin E enhances the absorption of vitamin A and increases the amount of vitamin A stored in the liver.

In the intestine, a small portion (less than 10%) of the retinol from food is oxidized first to retinal and then to retinoic acid, which is readily absorbed. Retinoic acid binds with albumin which increases its solubility in blood.

The chylomicrons (CM) containing vitamin A are removed from the circulation by liver after being slightly changed by the enzyme lipoprotein lipase. It is stored in the liver as droplets, as retinyl palmitate. When vitamin A is needed, it is released as retinol and carried on retinol-binding protein (RBP) which is synthesized in the liver. In the plasma this complex binds with *trans*-thyretin (TTR) and is transported to the tissues. Retinol–RBP is transported to hepatic stellar cells. Stellate cells may secrete directly into the plasma to combine with *trans*-thyretin (Figure 7.2).

Vitamin A is picked up by the cell by specific proteins within the cell known as CRBP (Cellular retinol-binding protein) and CRABP (Cellular retinoic-acid-binding protein) which specifically bind retinol and retinoic acid.

Vitamin A undergoes many enzymatic transformations in the body. In addition, retinal is reversibly reduced to retinol and irreversibly converted to retinoic acid (Figure 7.2).

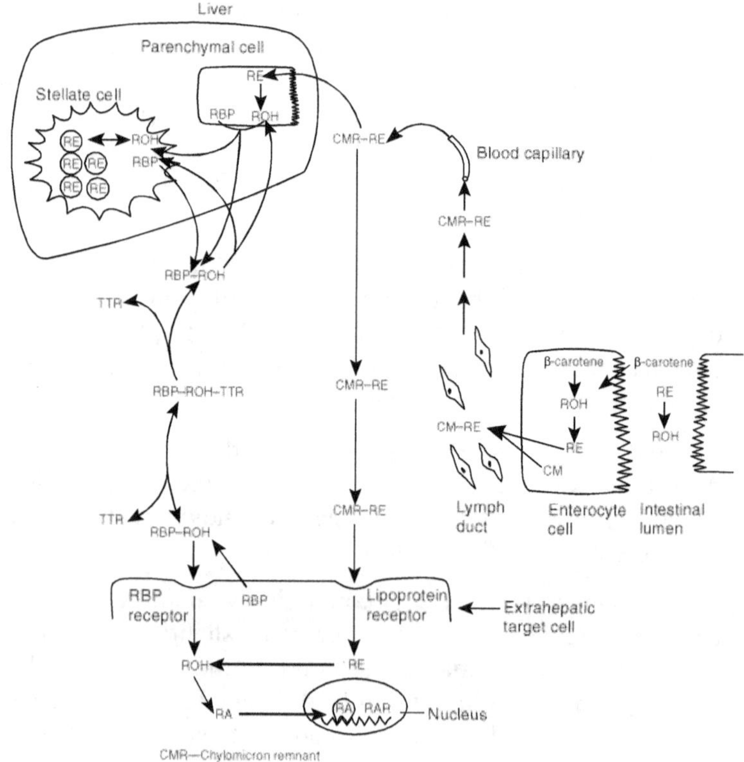

Figure 7.2 Summary of major pathways of retinoid transport

Circulating carotenoids, about 15–30% of which are β-carotene and the rest mostly non-provitamin carotenoids are carried by various classes of lipoproteins and the concentrations are directly related to dietary intake.

Vitamin A is depleted from the liver at a relatively low net rate of about 0.5% per day but within the body it is in a highly dynamic state. The half-life of RBP bound to *trans*-thyretin is about 11 hours, so it is one of the more rapidly turned over transport protein.

ASSESSMENT OF VITAMIN A STATUS

Liver reserves Vitamin A concentration in the liver is considered to be 0.07 µmol/g (20 mg/g) in both sexes.

Relative dose response (RDR) and the modified relative dose response (MRDR) is the indirect assessment of liver reserves made from a plasma assay. The basis for such tests is the finding that in the presence of vitamin A deficiency with diminished liver stores, apoRBP accumulates in the liver to several times its normal concentration. In the RDR, after baseline serum retinol has been measured, 450 µg retinyl palmitate is given orally. Some of this is taken up by the liver and combines with some of the excess apoRBP there and is released into the circulation in proportion to the pre-existing deficiency. Hence, this causes a rise in serum retinol, which is sampled again after about 5 hours. The result of the test is expressed as the difference between the two serum retinol values divided by the final value expressed as a percentage. Results of 50–70% indicate marginal liver stores, >5.50% indicates a deficiency.

Serum retinol Serum retinol level falls only when the liver stores have been exhausted and clinical signs of deficiency are also manifested. Even so, it is used to measure deficiency.

RECOMMENDED DIETARY ALLOWANCE

The RDA for vitamin A is given in the following table.

Group	Retinol μg	β-carotene μg
Man	600	2400
Woman	600	2400
Pregnant woman	600	2400
Lactation	950	3800
Infants	350	1400
Children		
1–6 years	400	1600
7–9 years	600	2400
Adolescents	600	2400

SOURCES

In animal foods, vitamin A is present in the form of retinol and its allied organic compounds which are highly bioavailable.

Good sources are agathi, amaranth, spinach, carrot, pumpkin, tomato, liver, egg yolk, milk, etc. Fruits and vegetables contain vitamin A only in the form of β-carotene.

VITAMIN A DEFICIENCY

Xerophthalmia is the main disease caused by Vitamin A deficiency, and is still a major cause of blindness in young children despite intensive prevention programmes. The parts of the world most seriously affected include South and East Asia, and some countries of Africa, Latin America and the Near East.

It has been estimated (WHO, 1991) that about 6–7 million new cases of xerophthalmia occur every year with about one in ten suffering corneal damage of these, 60% are dead within

1 year and of the survivors, 25% remain totally blind and 50–60% are partially blind.

Epidemiology

Xerophthalmia mainly affects children under the age of four years whose diet has been grossly inadequate for a long time. The severity of the eye lesions is in general inversely proportional to age, the blinding corneal changes targeting young children. Infections precipitate the condition (Figure 7.3).

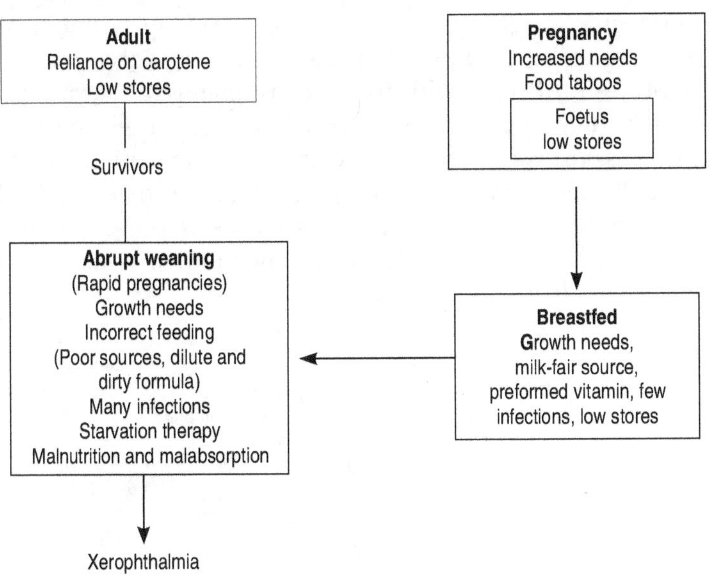

Figure 7.3 The xerophthalmia cycle

Manifestation of Xerophthalmia

Ocular manifestation Impaired dark adaptation is the first sign of xerophthalmia, most usually measured by dark adaptometry but also by rod scotometry. Changes in the fundus consisting of yellowish spots in the peripheral fundus, are visible on ophthalmoscopy.

The loss of goblet cells and early keratinizing changes in the bulbar conjunctiva, known as conjunctival xerosis and the more advanced stages of the same process called Bitot's spot, usually precede changes in the cornea. However, in very young children advanced corneal damage known as keratomalacia occurs without any obvious changes to the conjunctiva. Bitots spot consists of a heaping up of keratinized epithelial cells, most commonly occurring on the temporal aspect of the bulbar conjunctiva. It usually takes the form of a small plaque of a silvery grey hue with a foamy surface. The subjects are preschool children, and there is a prompt response to vitamin A. As deficiency progresses, the cornea becomes hazy but this stage appears to be of short duration as the corneal stroma rapidly becomes infiltrated with round cells, giving the cornea a milky bluish hue. The final stage consists of a unique pathological process known as colligative necrosis, in which the cornea virtually melts away to a varying degree in extent and depth. Frequently there is a total destruction of the eye with super added infection. Figure 7.4 shows the vitamin A deficiency that eventually leads to blindness.

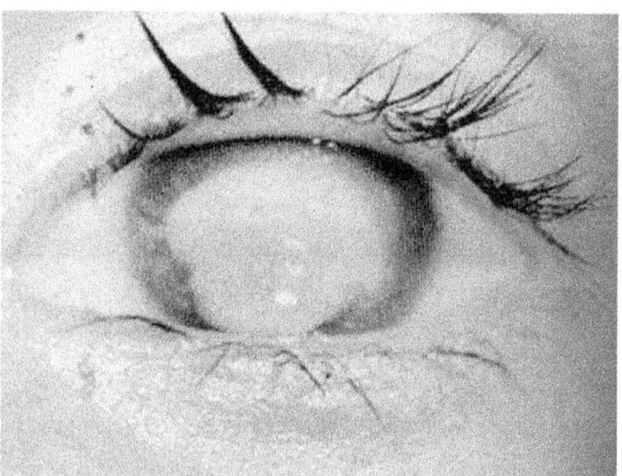

Figure 7.4 A vitamin A deficiency can eventually lead to blindness

Respiratory tract manifestation Ciliated epithelium in the nasal passages dries and the cilia are lost, thereby removing a barrier to entry of infections.

Gastrointestinal tract manifestation The salivary glands dry and the mouth becomes dry and cracked, open to invading microorganisms. The secretory function of the mucous membrane is diminished so that the tissue slough off, affecting digestion and absorption.

Genitourinary tract manifestation As epithetial tissue breaks down, problems such as urinary tract infections, calculi, and vaginal infections become more common.

Skin As the skin becomes dry and scaly small pustules or a hardened, pigmented papular eruption may appear around the hair follicles, a condition called follicular hyperkeratosis.

Tooth formation Because of lack of vitamin A, specific epithelial cells surrounding tooth buds in foetal gums tissue that become specialized cup-shaped organs called ameloblasts may not develop properly. These organs form the enamel structure of the developing tooth. Each cell carries out the task of producing and depositing minute prisms of enamel substance that eventually forms the erupted tooth.

Reproductive manifestation Deficiency causes sterility, testicular degeneration, abortions and malformed foetuses.

Treatment

Children between 1 and 6 years of age should receive 200,000 IU of retinyl palmitate by mouth immediately for any stage of xerophthalmia. Persistant vomiting or profuse diarrhoea necessitate the use of intramuscular injection, which must be of water-miscible form as the oily form does not leave the muscle. The dose is repeated the next day and again about 4 weeks later to provide liver stores against relapse. Children younger than one year or weighing less than 8 kg receive half the dosage.

Prevention

Periodic megadose vitamin A supplementation in the form of capsule, syrup or multiple dispenser 200,000 IU (66,000 μg) has been delivered at 4–6 monthly intervals either to high-risk preschool children or by inversal distribution to preschool children and lactating mothers. This programme is to save sight and lives on a large scale.

Food fortification Food is fortified with vitamin A, e.g. bread and milk in India.

Nutrition education Nutrition education, together with practical advice and help with growing cheap, nutritious vegetables and fruits in the home and school gardens could eradicate severe deficiency.

VITAMIN "A" TOXICITY

Acute toxicity following ingestion of several hundred thousand units of vitamin A has caused a rise in intracranial pressure, with vomiting, headache and papilloedema. With very large doses, drowsiness, repeated vomiting and skin exfoliation have occurred. Spontaneous recovery without residual damage usually follows on stopping the vitamin.

Chronic toxicity is more common, and is caused by the regular ingestion over a period of months or year of a dose usually in excess of 10 times the reference nutrient intake. Signs and symptoms occur insidiously, and the syndrome of headache, loss of hair, dry and itchy skin, and bone and joint pains are seen. Cessation of the vitamin is usually followed gradually by complete recovery, but in some cases liver damage, bone and muscle pain and impaired vision persist.

Congenital malformation in the offspring of mothers receiving large doses of vitamin A during the organogenetic period *in utero* has also been reported in animals.

CLINICAL USES OF VITAMIN A

Vitamin A and acne Vitamin A has been used to treat acne. The topical use of retinoic acid has been effective in some types of acne. However it should be used only under strict medical supervision and not used by pregnant women and those likely to become pregnant because it may cause birth defects.

Retinoid and cancer Vitamin A and especially carotenes function in preventing the proliferation or progression of cancer. Dietary intake of carotenoids is inversely related to the development of lung cancer. This may be due to their ability to act as antioxidants.

Carotene Carotenes are widely distributed in nature and all tissues capable of photosynthesis contain carotenes. Only a handful of carotenes have vitamin A activity. β-carotene is the most important form. β-carotene is made up of two vitamin A molecules while the other precursors yield only one molecule of vitamin A. Conversion occurs primarily in the intestinal wall and a little in the liver and lungs.

VITAMIN D (CALCIFEROL)

Vitamin D is the generic term for two molecules. Ergocalciferol (vitamin D_2) was obtained by irradiating the plant sterol ergosterol, which for many years was the major form of vitamin D used for the prevention and treatment of rickets. Cholecalciferol (vitamin D_3) is the major form of vitamin D in nature, but it can be made by irradiating 7-dehydrocholesterol and is more effective in preventing and curing rickets in chicks than vitamin D_2.

CHEMISTRY

The first stage of vitamin D_3 synthesis is the photoconversion of provitamin D_3 (7-dehydrocholesterol) to previtamin D_3. The amount of photoconversion depends on both the quantity and quality of the radiation reaching these layers of the epidermis.

Wavelengths of the order of 280–320 nm are required, with the maximum conversion occurring at 295 nm. After photoconversion, this previtamin can either undergo further photoconversion to tachysterol and lumisterol, or a heat-induced isomerization to vitamin D_3. The production of vitamin D_3 within the skin can take a period of several days. Vitamin D is stored in various fat depots throughout the body. The lipid nature of vitamin D and its metabolites limits their concentration within the circulation but a specific vitamin D transport protein (an α_2-globulin) binds a number of metabolites.

Conversion of vitamin D_3 to 25-hydroxy D_3 occurs in the liver. The enzyme 25-hydroxylase is found primarily in the microsomes and requires NADPH, molecular oxygen, magnesium ions and a cytosolic component. The circulating 25-hydroxy D_3 is further hydroxylated by one of two enzymes: 1α-hydroxylase which involves NADPH and incorporates molecular oxygen into the 1α-position of $25(OH)D_3$; and 24α-hydroxylase which incorporates the oxygen molecule into the 24α-position of $25(OH)D_3$ to yield $24, 25(OH)_2D_3$.

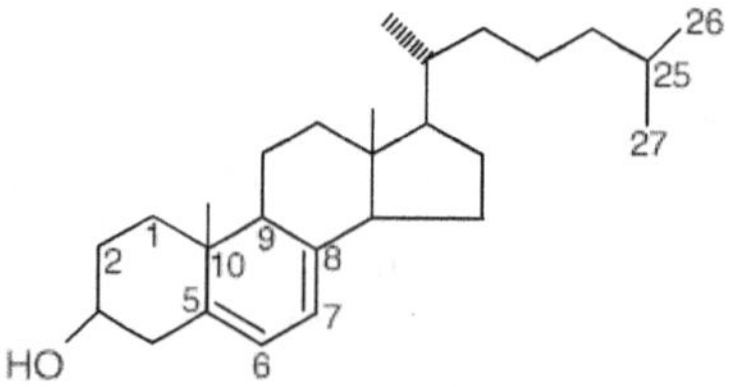

7-dehydrocholesterol
(Provitamin D_3)

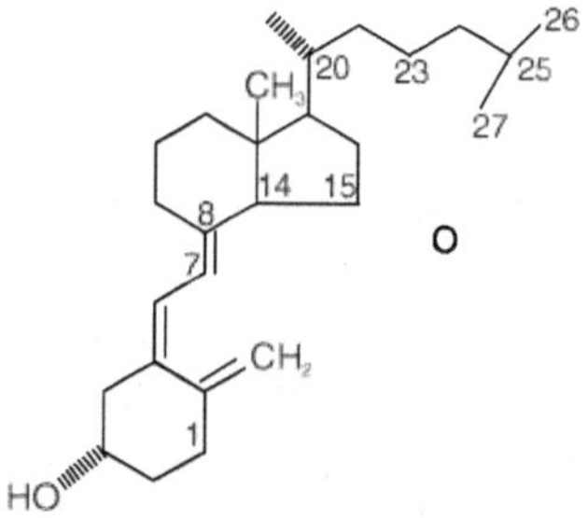

Vitamin D_3

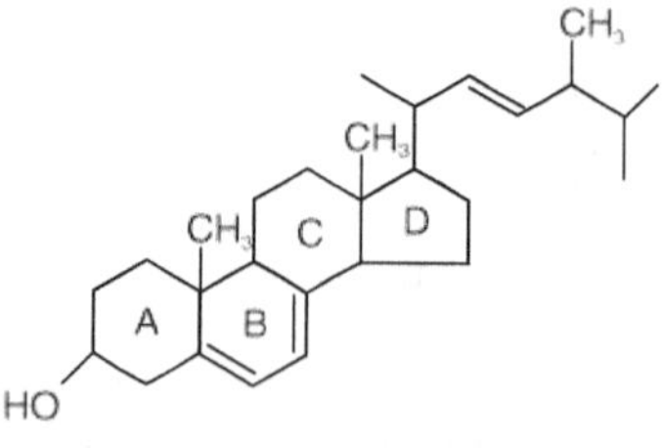

Ergosterol
(Provitamin D_2)

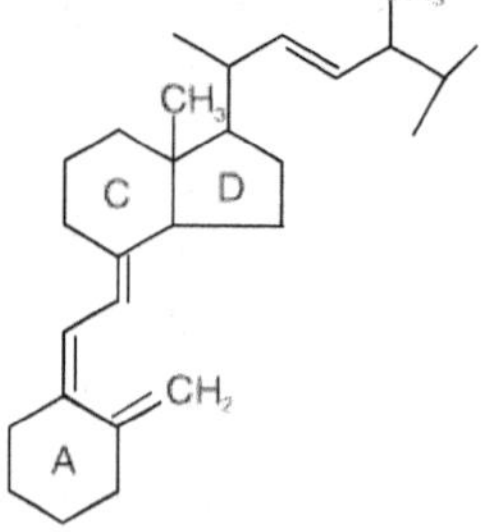

Vitamin D_2

Vitamin D transported by general circulation in combination with DBP (Vitamin D binding protein) to liver where, the enzyme 25-hydroxylase converts it to $25(OH)D_3$. Further conversion to the biologically active metabolite 1, $25(OH)_2D_3$ or $24,25(OH)_2D_3$ occurs in the kidney (Figure 7.5).

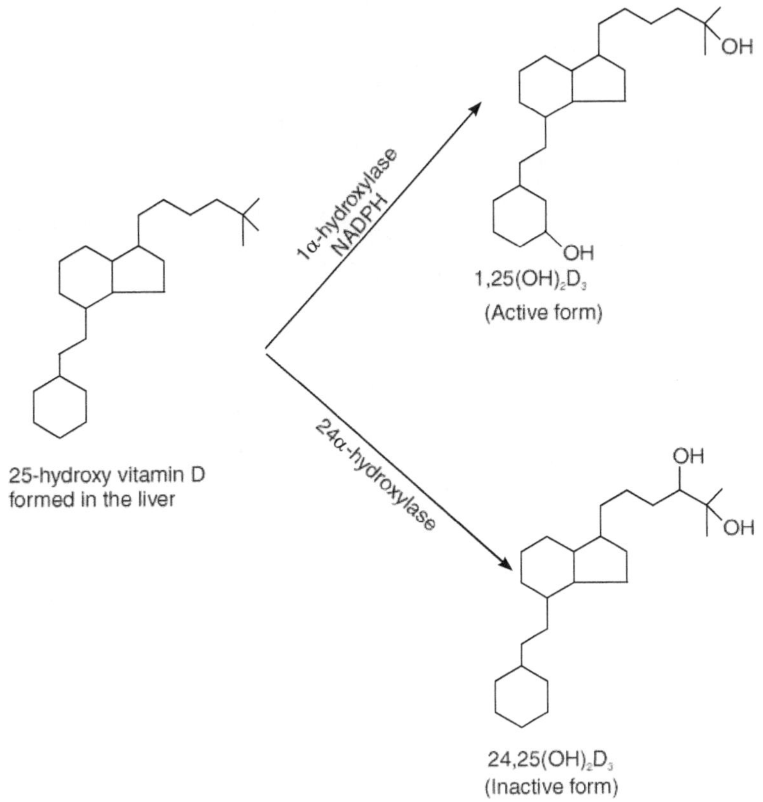

Figure 7.5 Formation of the active form and inactive form of vitamin D in kidney

During the continual exposure to the sun, pre-D_3 also photoisomerizes to lumisterols and tachysterols which are photoproducts that are biologically inert. DBP has no affinity for lumisterol and minimal affinity for tachysterol, the translocation of these photoisomers into the circulation is negligible and these photoproducts are sloughed off during the natural turnover of the skin or as soon as pre-D$_3$ stores are depleted. Exposure of lumisterol and tachysterol to ultraviolet radiation will provoke these isomers to photoisomerize to pre-D$_3$ (Figure 7.6).

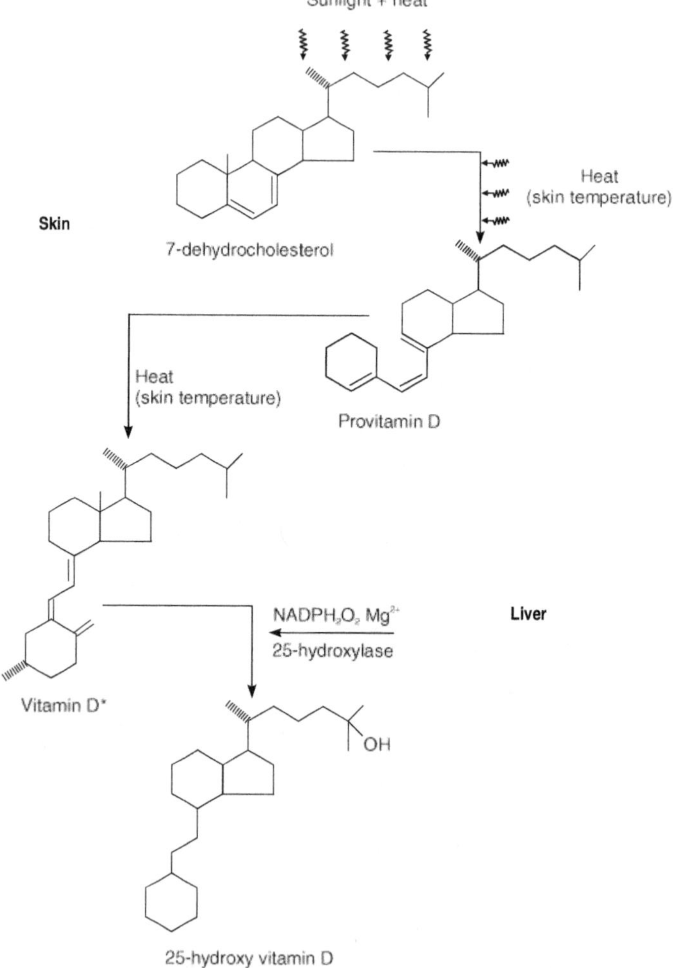

Figure 7.6   Formation of vitamin D in skin and liver

REGULATION OF PREVITAMIN D_3 SYNTHESIS IN HUMAN SKIN

Photochemical regulation Sunlight itself is responsible for regulating the total production of vitamin D_3 in human skin. Previtamin D_3 can thermally isomerize to vitamin D_3 or during exposure to sunlight, it can absorb ultraviolet radiation and isomerize to biologically inert isomers, lumisterol and tachysterol (Figure 7.7).

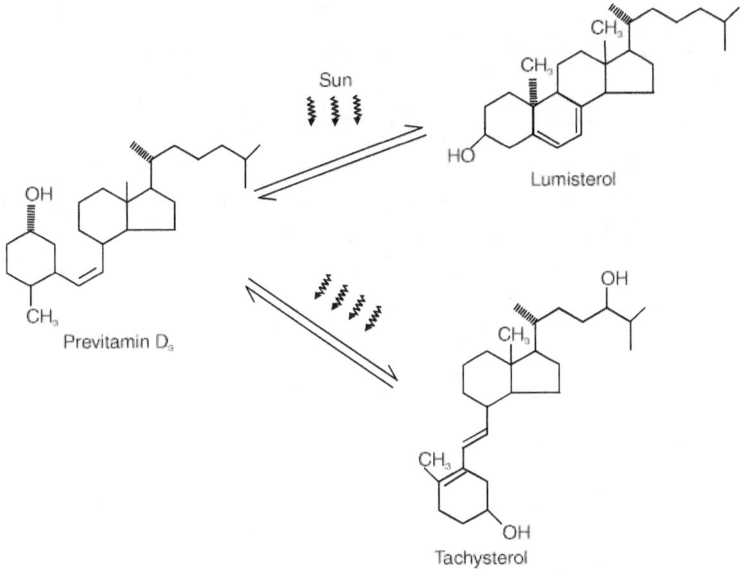

Figure 7.7 Photochemical regulation of previtamin D_3

Effect of age Amount of provitamin in skin decreases with age (in young adult the amount of provitamin D_3 in 1 cm^2 is approximately 0.8 µg for the epidermis and 0.15 to 0.5 µg for the dermis). Only 30% of that formed in young adults in produced in the elderly.

Sunscreen This reduces the amount of vitamin D_3 formed.

Control of synthesis Because 1, 25 (OH)$_2$ D_3 is the most biologically active form of cholecalciferol, its circulating concentration is tightly controlled. Although the kidney produces both dihydroxylated forms of the vitamin, the dominant form is determined by both the level of circulating parathyroid hormone

(PTH) and the body's vitamin D status. In a vitamin D-deficient state 1, $25(OH)_2D_3$ production is high, while that of 24, $25(OH)_2D_3$ is low. This is because $1,25(OH)_2D_3$ controls its own production in a negative feedback loop by suppressing the 1 α-hydroxylase enzyme. When vitamin D status is adequate, more $24,25(OH)_2D_3$ is produced. PTH, the production of which is stimulated by a fall in plasma calcium, increases the kidney's hydroxylase activity: when the body's vitamin D states is low the 1 α-hydroxylase is stimulated, but if vitamin D status is high then 24, $25(OH)_2D_3$ production is increased. 1, $25(OH)_2D_3$ also directly reduces PTH production within the parathyroid gland. Hypophosphataemia can also induce 1, $25(OH)_2D_3$ production. Thus a series of controls regulates the production of 1, $25(OH)_2D_3$.

Regulation by calcium   Increased dietary calcium decreases the synthesis of 1, $25(OH)_2D_3$ and increases the synthesis of $24,25(OH)_2D_3$. As the diet contains less of calcium, the synthesis and secretion of calcium-mobilizing hormone 1,25 DHCC is increased. At normal serum concentration, both $1,25(OH)_2D_3$ and $24,25(OH)_2D_3$ are made in equal amounts.

Regulation by PTH   PTH monitors the serum calcium concentration. PTH stimulates, synthesis of $1,25(OH)_2D_3$ in response to hypocalcaemia and shuts synthesis of $24,25(OH)_2D_3$.

Regulation by phosphate   Severe hypophosphataemia stimulates synthesis of $1,25(OH)_2D_3$ which accumulates even in the absence of PTH.

MECHANISM OF ACTION

Being lipid-soluble, vitamin D and its metabolites readily pass through the cell membranes to interact with a specific receptor. This receptor binds $1,25(OH)_2D_3$ more readily than $24,25(OH)_2D_3$ and $25(OH)_2D_3$ is least actively bound. In the blood, however, the vitamin D binding protein binds $24,25(OH)_2D_3$ and $25(OH)_2D_3$ more avidly than $1,25(OH)_2D_3$, so the $1,25(OH)_2D_3$ is readily concentrated within the cell. The receptor, with its bound $1,25(OH)_2D_3$ then translocates to the nucleus where it binds to the DNA of specific responsive genes. Special loops in the receptor common to all steroid hormone receptors and known as "zincfingers" for their zinc content, enable the receptor to interdigitate with the helical structure of DNA. Once bound to the DNA, the receptor induces messenger RNA production for the specific protein or peptide which is controlled by $1,25(OH)_2D_3$.

ACTION OF $1,25(OH)_2D_3$

Vitamin D maintains plasma calcium by stimulating intestinal calcium absorption by the small intestine and by increasing the resorption from bone.

Intestinal Calcium Absorption

$1,25(OH)_2D_3$ stimulates calcium transport across the intestinal cells by inducing the production of a calcium-binding protein (CBP) within the villus cells through the normal process of receptor binding, DNA interaction and mRNA production. The resulting CBP maintains the usually extremely low concentration of cytosolic calcium within the cell. Calcium diffuses into the villus cell passively, but calcium transfer into the blood seems to involve classic ion transport mechanisms. $1,25(OH)_2D_3$ also promotes cell maturation within the intestine. Intestinal mucosal cells have a short half-life and the length of the villus in vitamin D-deficient rats is only some 70–80% of that in normal rats.

Bone Resorption

In a mature adult in normal calcium balance, a demand in excess of the supply of calcium is met by skeletal calcium because appropriate concentrations of calcium are important in vital functions such as nerve and muscle activity. As plasma calcium falls, there is a rise in PTH which in turn increases $1,25(OH)_2D_3$ synthesis. This then acts on the bone-forming cells, the osteoblasts, which produce factors that stimulate the activity of the bone-removing cells, the osteoclasts. $1,25(OH)_2D_3$ also promotes bone resorption by increasing the formation of new osteoclasts. $1,25(OH)_2D_3$ therefore leads to an increase in both the number and activity of osteoclasts within the bone and these remove the bone matrix to release the bound calcium.

Growth-plate Mineralization and Bone Formation

$1,25(OH)_2D_3$ increases bone formation by providing sufficient calcium within the body to allow calcification to occur. $1,25(OH)_2D_3$ also stimulates the production of osteocalcium. This protein is formed by osteoblasts and binds up to four calcium molecules and is found exclusively in the bone $1,25(OH)_2D_3$ also affects other processes involved in bone formation, including alkaline phosphatase activity and collagen synthesis. Within the growth plate, chondrocytes proliferate, differentiate and finally become calcified to form bone. This process results in longitudinal growth. Vitamin D metabolites may affect the mineralization of the skeleton by altering chondrocyte differentiation.

Other Actions of $1,25(OH)_2D_3$

1. Plays a role in the process of cell proliferation and maturation.

2. Children with rickets often have abnormalities associated with immunohaematopoietic system, with increased frequency of infections, impaired neutrophil phagocytosis, anaemia and decreased bone marrow cellularity. These are corrected by vitamin D.

3. Plays a role in the control of cell maturation in the skin.

4. $1,25(OH)_2D_3$ may also have a role in the treatment of leukemia and lung and colon cancers, where it may inhibit proliferation.

5. Vitamin D helps deposition of crystalline $1,25(OH)_2D_3$ in the bone by supplying Ca and P in proper concentration to the bony matrix. A rate of 10% absorption without vitamin D increases to 33% in its presence.

6. Vitamin D increases the rate of reabsorption of phosphorus from the kidney tubules.

7. In the intestines, calcitriol stimulates the synthesis of a phosphate-binding protein which increases absorption of phosphorus.

8. It regulates the level of the enzyme alkaline phosphatase in the serum. Phosphatase helps in the deposition of $Ca_3(PO_4)_2$ in bones and teeth.

9. Vitamin D is necessary for the secretion of insulin from β-cells of the pancreas. Insulin secretion was found to be decreased in vitamin D-deficient rats.

ASSAY

1. Vitamin D status is assessed by measuring $25(OH)_2D_3$.

2. Circulating levels of $1,25(OH)_2D_3$ are measured with lipid extracts of plasma or serum.

The normal circulating concentration in 38–144 pmol/L or (16–60 pg/ml).

The plasma concentration of $1,25(OH)_2D_3$ is 1000 times greater than $1,25(OH)_2D_3$.

ABSORPTION AND METABOLISM

Dietary vitamin D is absorbed in the upper part of the small intestine. Both infants and adults absorb 80% of intake. The same factors that aid in or depress fat absorption also enhance or depress absorption of vitamin D.

Once absorbed, it is incorporated into chylomicrons. The summary of absorption and metabolism of vitamin D is given in Figure 7.8.

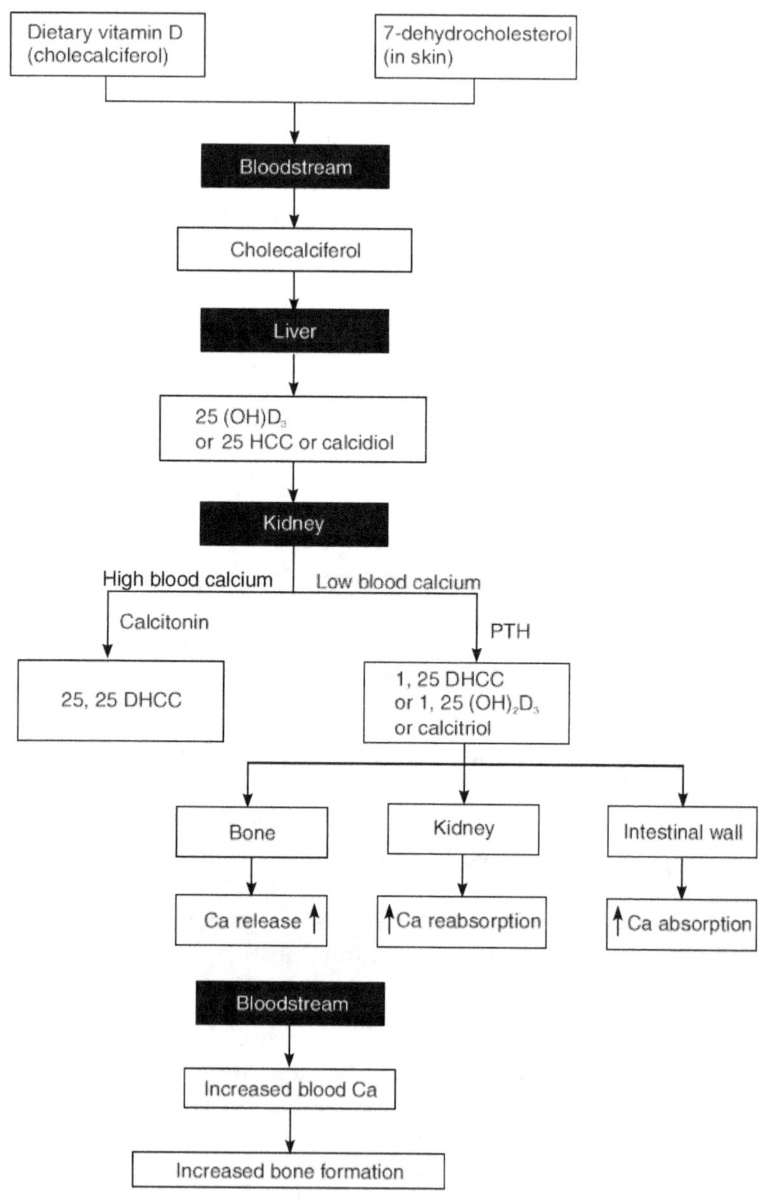

Figure 7.8   Absorption and metabolism of vitamin D

SOURCES

Vitamin D is available to the body by two separate pathways.

1. Skin

2. Diet

Skin The precursor 7-dehydrocholesterol when exposed to UV light (ray) from the sun is converted into cholecalciferol (D_3). The amount of vitamin D produced through irradiation of the precursor in the skin is influenced by the amount of UV light a person is exposed to.

During summer, individuals may get enough UV light but UV light cannot penetrate fog, clouds, smoke, window, glasses or clothing. The pigment melanin also protects against the overproduction of vitamin D.

Diet Vitamin D is present mainly in foods of animal origin and only very minimally in plant foods. The natural sources are egg yolk, liver, milk, butter and fish and fish liver oils. A vegetarian diet may not contain vitamin D unless milk is taken. The vitamin D content of food depends upon exposure of animals to sunlight. Vitamin D is usually added to commercial baby food and milk powder.

Important sources	μg/100 g
Fish liver oils	30–10,000
Fatty fish	5–30
Egg yolk	3–4
Butter	0.5–1.5
Milk	0.05–0.1
Milk powder	0.4–0.6

REQUIREMENTS

The requirements of vitamin D for various age groups are as follows:

Infants and preschool children—10 µg

Older children and adults—5–7.5 µg

During lactation and pregnancy—10 µg

In tropical climate, half the amount will be adequate if subjects are exposed to direct sunlight for at least some hours daily.

DEFICIENCY

Deficiency of vitamin D occurs with

1. inadequate exposure to sunlight

2. inadequate intake

3. drugs that hinder vitamin D absorption and action

4. altered metabolism due to damaged liver and kidney

Deficiency of vitamin D leads to diminished absorption of calcium and consequent low plasma calcium which may produce tetany. Low plasma calcium stimulates the parathyroid to mobilize bone calcium for raising the plasma calcium levels. The bones are thus depleted of calcium salts.

Effects of Vitamin D deficiency

A deficiency of vitamin D leads to

1. Inadequate absorption of Ca and P from the intestinal tract and

2. Faulty mineralization of bones and teeth. The soft bones are unable to withstand the stress of weight resulting in skeletal malformations.

RICKETS

It is known as English disease and is a disease of defective bone formation that manifests itself in many ways as poorly calcified bones are not able to perform their function.

Aetiology

1. Use of purdah as a result of which the individual is not exposed to light.

2. Smoky, industrial areas of cities where buildings are over-crowded shutting out sunlight and having poor sanitation.

3. Dark-skinned people moving from tropics to temperate zones due to lack of exposure to sunlight.

4. It is a disease of the poor. Milk (provides calcium), butter, and egg, the only common foods containing a satisfactory dietary source of vitamin D is too expensive for poor families.

5. Premature infants start life deficient in vitamin D and bone salts Ca and P, since a large amount of foetal skeleton is mineralized during the last trimester of pregnancy.

Clinical Features

The following are the characteristic clinical features seen in infants with rickets (Figure 7.9).

1. The infant is restless, fretful and pale.

2. Flabby and toneless muscles which make limbs assume unnatural postures (acrobatic rickets).

3. Excessive sweating of head.

4. Abdomen distended as a result of flabby abdominal muscles.

5. Teeth erupt late.

6. Failure to stand, crawl, walk at normal ages.

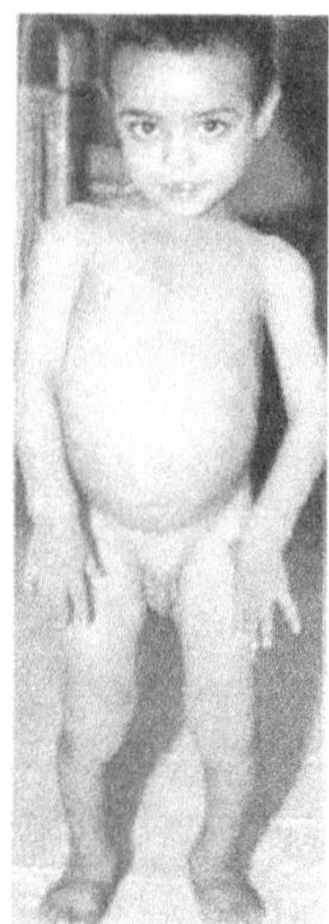

Figure 7.9   The bowed legs of rickets

7. Bony changes most characteristic are

   i. Earliest change in craniotabes which are small round unossified areas in the skull yielding to pressure of the finger and gives a crackling feeling.

   ii. The ends of long bones become enlarged causing difficulties in movement (epiphyseal enlargement).

   iii. Ribs develop irregularly spaced areas of swelling that look like beads (rachiatic rosary).

   iv. Bossing of frontal and parietal bones.

   v. Chest becomes concave and causes overcrowding in the chest cavity (pigeon chest).

   vi. The fontanel (point at which the two hemispheres of the skull merge; the juncture is open at birth but closes by one year) which is the opening on top of the skull of a very young infant, fails to close in early life so it causes rapid enlargement of the head.

vii. Since bones are soft due to non-depostion of Ca salts, they are easily bent by the weight of the body.

If rickets continues during second and third year of life serious bone deformities such as bow legs, knock knees and deformities of the spine and pelvis develop.

Treatment

Supplement of vitamin D (1000–5000 IU) per day and a good amount of calcium is supplied. The hygienic environment of the child must be improved. Unnecessary clothing should be removed and the child should be allowed to enjoy sunshine as much as possible.

Prophylaxis   The amount of vitamin D required to prevent rickets in infants and children will depend upon their exposure to sunlight and diet. A total daily intake of 10 µg of vitamin D provided by 1 tsp of cod liver oil is a protection against rickets.

Cure   As therapy for rickets oral vitamin D can be given. Natural vitamin D_3 can also be given as cod or shark liver oil. Oral administration is quite effective. So vitamin D injections are not required.

OSTEOMALACIA
(MALACIA MEANS SOFTENING)

Osteomalacia is similar to rickets but occurs in the older age group. It occurs due to

1.   Lack of dietary intake of vitamin D or sunlight.

2.   Malabsoption of vitamin D.

3.   Excessive secretion of milk during lactation if it is prolonged.

4.   Excessive demand as in successive pregnancies.

5.   Impaired 25, hydroxylation of vitamin D in the liver.

6. Impaired 1, hydroxylation in the kidney with chromic kidney disease or PTH deficiency.

In osteomalacia the bony mineral content decreases due to lack of deposition of $Ca_3(PO_4)_2$. There is also bony resorption of mineral due to increased PTH activity following low plasma calcium.

Ostcemalacia occurs when there is less calcium in the body than required. The reason for low calcium levels are

1. a poor diet and a low intake of milk.

2. calcium in the diet is not well absorbed because of clothing/purdah which shields them from exposure to sunlight leading to vitamin D deficiency.

3. repeated pregnancies and prolonged lactation further deplete body Ca stores.

The pigment of dark-skinned people is a protection from excessive vitamin D_3 production in the tropical sun but when they migrate to the temperate zone and have decreased exposure to sunlight and wear protective clothes, eat whole grain cereals with high phytates, it interferes with calcium absorbed.

Certain drugs like phenobarbitone and phenytoin which are used for treatment of epileptics inhibit Ca transport and bone mineral mobilization and metabolize vitamin D into inactive products and enhance excretion through bile.

Osteomalacia may also occur in post gastrectomy patients.

Osteomalacia manifests itself as skeletal pain, bony tenderness and fractures and bony deformities of the pelvis, legs, ribs and lower vertebrae. Having deformed pelvis, a woman cannot have a normal delivery.

Treatment

500–10,000 IU/day of vitamin D and consumption of vitamin D-enriched milk.

DENTAL CARIES

A deficiency of vitamin D may lead to delayed dentine formation and to malformation of the tooth and dental caries.

Studies have shown that children who have received increased amounts of milk, meat, egg and other vitamin D sources have few caries than others.

Hypoparathyroidism requires 1–2 µg vitamin D daily along with liberal milk intake to supply calcium. The blood calcium level should determine the dosage of vitamin D.

Unlike vitamin D, calcitriol acts directly on the intestine for calcium absorption and does not require action in the liver, PTH or kidney. Calcitriol raises plasma calcium in a few days.

VITAMIN D TOXICITY

The fat-soluble vitamin in contrast to those that are water-soluble is not rapidly excreted or metabolized and if taken in excessive amounts may accumulate in the body and produce undesirable toxic effects.

It rarely happens because of its presence in natural sources but it can occur in

1. patients treated with vitamin D for hypothyroidism

2. patients treated for osteoporosis and osteomalacia

3. children who are mistakingly given excess doses

In mild toxicity (2000–3000 IU) is associated with hypercalcaemia, are given.

For excess toxicity 1,50,000 IU for 10–15 days is given.

There is no evidence of toxicity form overexposure to the sun.

Toxic Symptoms

- sudden loss of appetite

- nausea and vomiting

- ☙ excessive thirst

- ☙ polyuria

- ☙ severe constipation/diarrhoea

- ☙ pain in head

- ☙ child is thin, irritable and depressed and may lead to meningitis

- ☙ hypercalcaemia

- ☙ hypercalciuria

- ☙ joint pains

- ☙ calcification of soft tissues such as lungs and kidney and bone fragility

Treatment

In mild cases withdraw vitamin D till calcium in serum falls, but in severe cases glucocorticoid, calcitonin must be given. Normal vitamin D level is 1.7–4.1 μg/dl.

VITAMIN E

The existence of vitamin E was recognized in 1922. In 1922 a fat-soluble dietary constituent was found to be essential for the prevention of foetal death and sterility in rats. This was originally called "factor X" and antisterility factor but was later named as vitamin E. When finally isolated from wheat germ oil in 1936, vitamin E was called tocopherol from the Greek word *tokes* and *pherein*, meaning to bring forth children.

CHEMISTRY

Pure vitamin E is odourless and colourless. It is soluble in organic solvents and is insoluble in water. There are eight naturally occurring vitamin E compounds which are synthesized by plants.

The chemical structure of naturally occurring tocopherol and tocotrienol is given above. The tocotrienols differ from the tocopherols in that the 16-carbon phytol chain contains three unsaturated double bonds.

R_1 R_2 R_3 are the number and position of the methyl groups on the chromanol ring which denotes the different homologues.

FUNCTION

It is a highly effective antioxidant, readily donating the hydrogen from the hydroxyl group on the ring structure to free radicals. Vitamin E has a major biological role in protecting polyunsaturated fats and other components of the cell membranes from oxidation by free radicals and is therefore located within the phospholipid bilayer of the cell membrane.

Free Radicals

The production of free radicals which can initiate damage to biological material occurs during normal aerobic metabolism.

Free radicals entering the body from the atmosphere or formed during metabolism cause oxidation of PUFA which destroys membranes. Activated oxygen species are formed during the stepwise reduction of oxygen to water and by secondary reactions with protons and transition metals such as copper and iron. For example, the superoxide anion (O_2^-) is produced in many cell redox systems.

Lipid Peroxidation and Vitamin E

Polyunsaturated fatty acids (PUFA : H) are major constituents of cell membranes. They are particularly susceptible to free-radical-mediated oxidation. Thus the process of lipid peroxidation can lead to disturbances in membrane structure and function. The process is initiated by a free radical such as OH extracting a hydrogen from PUFA : H, with the formation of PUFA radical (PUFA). This is followed by the rearrangement of the double bond to form a conjugated diene which then combines with oxygen to produce a peroxyl radical (PUFAOO) which in turn reacts with more PUFA : H to form a hydroperoxide (PUFA : OOH) and another PUFA. The reaction is now self-propagating.

$$PUFA:H^\bullet \rightarrow PUFA^\bullet$$

$$PUFA^\bullet + O_2 \rightarrow PUFA:OO^\bullet$$

$$PUFA:OO^\bullet + PUFA:H \rightarrow PUFA:OOH + PUFA^\bullet$$

Moreover in the presence of iron or copper PUFA : OOH can undergo a fission of the double bonds and one electron reduction to form more free radicals including the highly reactive $OH^\bullet$.

Such auto-oxidation will continue unless all the free radicals are scavenged by antioxidants. Therefore a major biological role for vitamin E is to break the chain of events leading to the formation of PUFA : OOH by donating the hydrogen atom of the hydroxy group on its chromanol ring to the free radical to form a stable compound.

If vitamin E fails to prevent the formation of PUFA : OOH in the cell membrane, PUFA : OOH can be released from phospholipid by phospholipase A_2 and then degraded by selenium-containing glutathione peroxidase in the cell cytoplasm. Thus the antioxidant activities of vitamin E and selenium through glutathione peroxidase are closely related. Other antioxidant enzymes include superoxide dismutase, glucose 6-phosphate dehydrogenase, and compounds such as carotenoids, and ascorbic acid also have antioxidant properties (Figure 7.10).

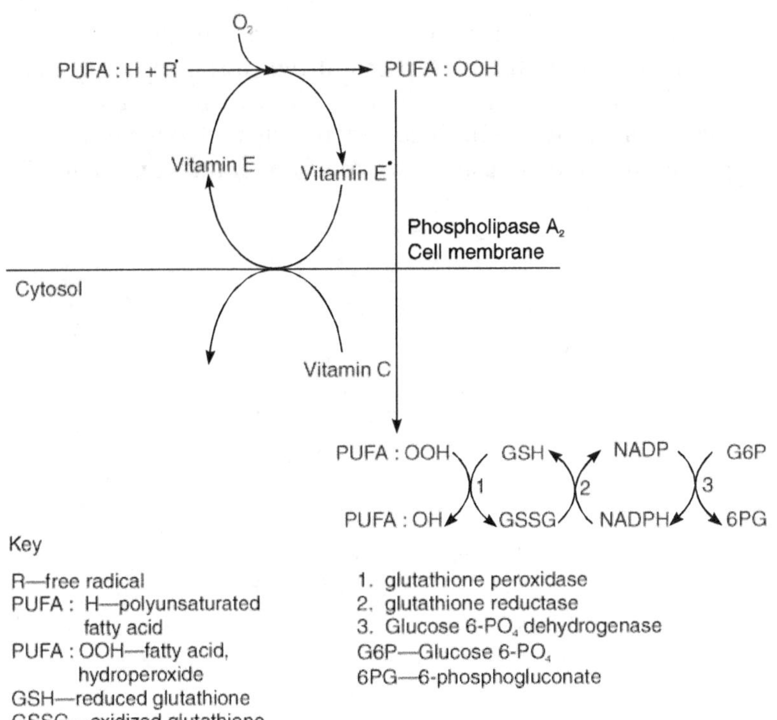

Figure 7.10 Some of the major antioxidant defence mechanisms within the cell

Other Roles of Vitamin E

Vitamin E may also play important roles in other biological processes. These include:

1. Structural role in the maintenance of cell membrane integrity

2. Anti-inflammatory function

3. DNA synthesis

4. Stimulation of immune response

5. Protecting vitamin A, C and unsaturated fatty acids

The antioxidant property of vitamin E also prevents the formation of lipofuscin, a pigment that forms in the uterus and may be involved in the reabsorption of foetal tissue during miscarriage. Lipofuscin is considered characteristic of the ageing process and shows up as brown spots under the skin (Figure 7.11).

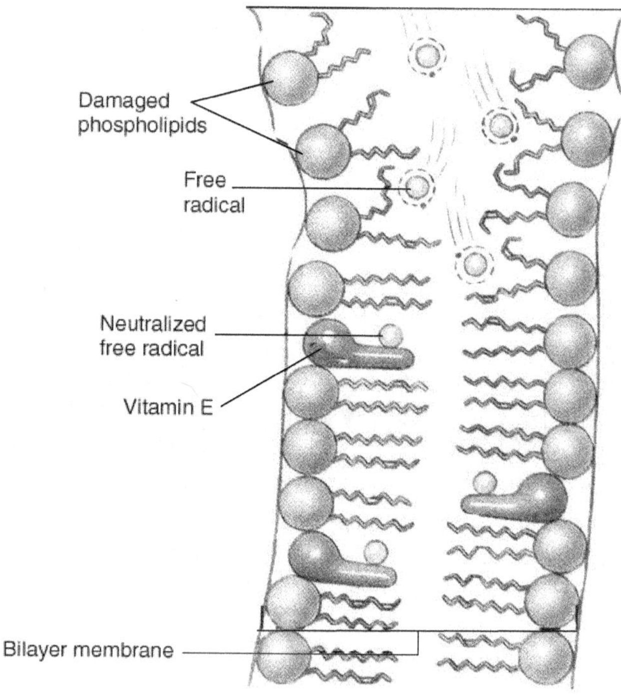

Figure 7.11   Vitamin E helps stop free radical chain reactions

ABSORPTION AND METABOLISM

Tocopherols are absorbed from the gut in micelles whose formation depends on bile salts and pancreatic lipase. Maximum absorption occurs in the upper and middle third of the small intestine. Tocopherol uptake is enhanced by medium-chain triglycerides but inhibited by long-chain polyunsaturated fatty acids. Both α and γ-tocopherol are absorbed. Micelles containing tocopherol may passively diffuse through the brush border. It is released into the lymph within chylomicrons.

The resulting chylomicron remnant is taken up by the liver which secretes the α-form into the plasma in the newly formed VLDL but the α-form of tocopherol is secreted, into the bile and is excreted.

Tocopherols incorporated within lipoproteins(LDL) are taken up mainly by the low-density lipoprotein receptors on peripheral cells. They are transported within the cell and incorporated into the cell membrane.

Tocopherol equivalents (TE):

1 μg tocopherol   = I IU α-tocopherol

= 1 tocopherol equivalent

Vitamin E is stored in the adipose tissue, muscle and liver.

FOOD SOURCES

Vegetable oils, nuts and whole grams are the richest sources. Vitamin E requirement is linked to that of essential fatty acids RDA suggested is 0.8 μg/g of essential fatty acids.

EVALUATION OF VITAMIN E STATUS

Serum vitamin E level   Serum levels below 0.6 μg/dl suggest vitamin E deficiency while levels about 3.5 μg/dl is associated with toxicity.

Erythrocyte haemolysis test   Vitamin E offers protection against the peroxide-induced erythrocyte membrane damage responsible for haemolysis. Hence, the resistance of the RBC to haemolysis is an indirect measure of vitamin E status.

DEFICIENCY

In humans, vitamin E deficiency is seen in premature infants and adults with defects in fat absorption.

The most common symptom observed in these cases in an increase in haemolysis of the RBC.

In the only study of an induced vitamin E deficiency in human men receiving an intake of 4 mg of tocopherol/day for 30 months showed a steady drop in plasma vitamin E to half the normal values. The plasma tocopherol levels fell to 1 µg/dl and was accompanied by an increase in haemolysis of red blood cells.

The anaemia that results from the breakdown of the red blood cells is seen in infants, and children with cystic fibrosis, have difficulty in absorbing lipids. For such children water-miscible form of the vitamin is given. This has been associated with increased platelet aggregation.

Vitamin E deficient infants are very sensitive to oxygen therapy which damages the retina of the eye and can cause permanent blindness. Vitamin E in doses of up to 100 mg/day protects against this condition. This condition is called retrolental fibroplasia. Retinopathy in premature infants can be prevented by large doses of vitamin E.

In animals nutritional muscular dystrophy, characterized by muscle weakness caused by fragmentation of the muscle fibres and the accumulation of fluid is seen among vitamin E deficient guinea pigs and rabbits.

Neurological disorders characterized by ataxia is seen in vitamin E deficient diets in some animal studies.

In several species of animals, vitamin E deficiency has also been associated with a decreased ability of T and B lymphocytes to proliferate as part of body's immune defences.

CLINICAL USE

Vitamin E and rheumatoid arthritis Clinical trials in humans have shown that dietary supplement with ω-3 fatty acids found in fish oil provide significant benefits. It reduces the levels of inflammation-causing cytokines, the proteins that cause joint swelling, pain and tenderness which is a characteristic of the disease.

TOXICITY

Vitamin E is well tolerated and acute and chronic toxicities are very rare. Relatively few side effects are seen in humans even at intakes as high as 3200 IU/day. Some cases have shown that it may cause breast soreness, emotional disorders, muscular weakness and gastrointestinal disorders.

THERAPEUTIC USES

As free radicals have been implicated in many diseases, vitamin E has been used as therapeutic agent. High doses of vitamin E may slow down the progression of Parkinson's disease and some neurological disorders.

An increased vitamin E level is suggested for smokers as tobacco smoke contains vast quantities of reactive free radicals.

VITAMIN K

A dietary derived coagulation factor was first described by Dam and Schonheyder in Denmark. They observed a bleeding disorder in chickens which was corrected by feeding a variety of foods but mostly alfalfa. Dam termed the active factor "koagulation-vitamin" in 1935. He along with Karrer and co-workers isolated the newest fat-soluble vitamin in 1939.

CHEMISTRY

Vitamin K exists in two forms; both are 2-methylnaphthoquinone rings with side chains. Vitamin K_1, or phylloquinone has a phytyl side chain and occurs only in plants. Vitamin K_2 belongs to a family of compounds called menaquinones.

Phylloquinone (K_1)
(plant derived, lipid-soluble)

Menaquinones (K_2)
(bacterially derived lipid-soluble)

Menadione (K_3)
(synthetic, water-soluble)

The biological activity of vitamin K is due to its ability to change between the oxidized forms (quinone and epoxide) and the reduced form (quinol) in the vitamin K cycle.

The conversion of quinol to the epoxide form is calatysed by an enzyme with epoxidase and carboxylase activity. The carboxylase converts glutamate (glu) residues of proteins to γ-carboxyglutamate (glu) residues. These proteins are termed vitamin-K-dependent or gla-proteins. The enzymes of the vitamin K cycle are found in rough endoplasmic reticular membranes of liver and bone. The gla-proteins are able to bind calcium ions very readily and it is this property that gives them their biologic activity (Figure 7.12).

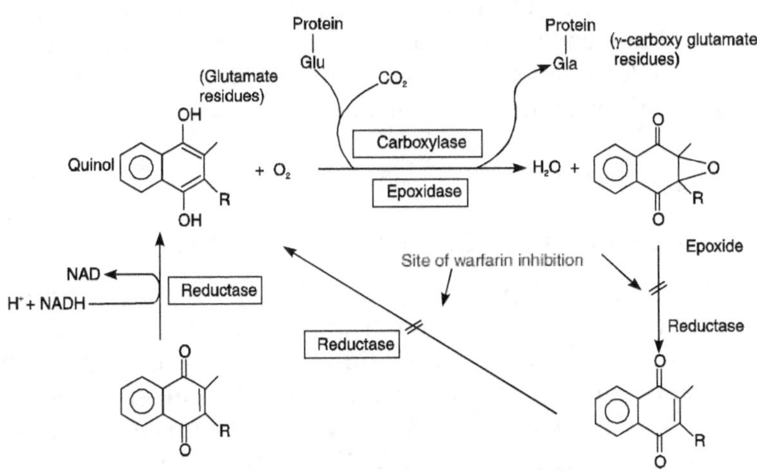

Figure 7.12 Vitamin K cycle

VITAMIN-K-DEPENDENT PROTEINS

Plasma Proteins

The first established vitamin-K-dependent proteins were the coagulation proteins, factors II (prothrombin) VII, IX and X. Their calcium binding property enable them to associate specifically with the acidic phospholipids on cell and platelet membranes which is necessary for the initiation and continuation of the coagulation cascade. Two further plasma vitamin-K-dependent proteins C and S inhibit coagulation by blocking the activation of factors V and VIII.

Bone Proteins

The main protein of mineralized tissue (bone and dentine) is osteocalcin which is produced by osteoblast and odontoblast cells respectively. A further vitamin-K-dependent protein which is found in both mineralized tissue and cartilaginous tissue is matrix protein. Osteocalcin and matrix protein bind hydroxyapatite and help in bone mineralization.

Other proteins   Other vitamin-K-dependent proteins have been found in many other tissues like the kidney, placenta, pancreas, spleen, lung, testis and atherosclerotic plaques. In the brain they are vital for brain development.

PHYSIOLOGICAL FUNCTION

Regulation of Clotting Protein Synthesis

The vitamin-K-dependent coagulation proteins are proenzymes that are converted to serine hydrolases during coagulation. All require calcium for activation. This is mediated by their carboxyglutamate residues. The vitamin-K-dependent procoagulant factors (II, VII, IX and X) form the core of the proteolytic cascade leading to fibrin formation in haemostasis. Protein C is an anticoagulant and serves as a brake on the speed of the intrinsic cascade through a feedback loop involving thrombin. While thrombin is producing fibrin, it is also activating protein C which in turn inactivates factors V and VIII thus slowing down the process. Protein S enhances the activity of protein C in the presence of phospholipid.

Thus the coagulation cascade is tightly regulated by both procoagulants and anticoagulants, many of which are dependent on vitamin K.

Absorption, Distribution and Metabolism

The absorption of phylloquinone and menaquinones requires bile and pancreatic juice. Dietary vitamin K is absorbed in the small intestine incorporated into the chylomicrons and appears in the lymph. 50–80% of the vitamin K is absorbed. It circulates in the plasma in VLDL. Any condition which impairs fat absorption such as biliary obstruction, pancreatic insufficiency and coeliac disease will decrease absorption.

The liver provides the major store of vitamin K of which around 90% are menaquinones.

After recycling many times over, vitamin K is metabolized to a variety of water-soluble, bile acid-conjugated products which are excreted in the urine and faeces.

SOURCES

Vitamin K_1 occurs in plants in the quinone form and is particularly abundant in green leafy vegetables such as spinach, (300–400 μg/100 g) cabbage, etc. Margarine and soya oil are rich sources.

Cows milk contains a small quantity of vitamin K. The liver is the storage site for vitamins K_1, and K_2 and is a dietary source of the vitamin.

Non-dietary source includes bacteria in the intestine and colour which synthesize menaquinones.

1 μg/kg/day of vitamin K has been the suggested RDA for adults and for infants a higher value of 2 μg/kg body weight is recommended since their liver stores are lower and their gut flora is poorly developed.

ASSAY

Bleeding tendency is estimated. This, can be identified by bleeding from the nose or mouth, echymoses in the groin, around collar line or in the legs; haematurea and haematemesis are all signs of vitamin K deficiency.

A reduction in the prothrombin and other vitamin-K-dependent factors in the plasma are also estimated.

FACTORS AFFECTING VITAMIN K STATUS

Effect of diet Vitamin K deficient diets in humans has shown depressed plasma vitamin K levels.

Diets that enhance colonic fermentation (high in soluble non-starch polysaccharides and resistant starch) may increase menaquinone production. The absorption of vitamin K_2 from the colon and lower ileum is by passive uptake.

Effect of drugs and surgical interventions The classic anticoagulant drugs are dicoumarins (e.g. warfarin) which were first isolated from sweet clover in 1941. They block the recycling of vitamin K and result in bleeding.

Broad-spectrum antibiotics destroy menaquinone producing intestinal bacteria.

Biliary obstruction The secretion of bile salts is essential for the absorption of fats and fat-soluble vitamins and hence biliary obstruction causes vitamin K deficiency.

Malabsorption syndrome Depression of the vitamin-K-dependent coagulation factors is found in malabsorption syndromes and other gastrointestinal disorders. Patients with malabsorption should be treated with vitamin K orally in doses of 2.2 to 4.4 mmol per day.

Liver disease Patients with liver disease are unable to utilize vitamin K in the biosynthesis of vitamin-K-dependent clotting factors, usually as a result of destruction of the rough endoplasmic reticulum in the hepatocytes.

Megadoses of vitamin A and E Megadoses of the fat-soluble vitamins A and E are known to antagonize vitamin K. Hypervitaminosis A in the rat leads to hypoprothrombinaemia which can be corrected by vitamin K administration.

It was also observed that high vitamin E intakes in rats increased vitamin K requirement.

DISEASES ASSOCIATED WITH VITAMIN K

Haemorrhagic disease of the newborn Levels of vitamin K in human milk vary and appear inadequate for some babies who are solely breast fed. This syndrome, termed haemorrhagic disease of the newborn is characterized by spontaneous bruising or bleeding or intracranial haemorrhage, which can result in death.

Thrombosis-associated diseases   These include coronary heart disease (CHD) and veno-occlusive disease (VOD). It is considered that the hypercoagulable state that predispose to CHD and VOD may be contributed to by relatively high levels of functional vitamin-K-dependent proteins (Factors II, VII, IX and V).

Osteoporosis   There is a lower vitamin K status in patients with osteoporosis due to the vitamin-K-dependent proteins osteocalcin and bone matrix protein.

TOXICITY

Phylloquinone toxicity is very rare but menadione toxicity occurs.

The uncontrolled use of menadrone, the synthetic form of vitamin K resulted in an increase in a haemolytic type of anaemia, an accumulation of bilirubin in the blood, and a condition known as kernicterus in which bile pigment accumulates in the grey matter of the central nervous system. Kernicterus is characterized by mental retardation, jaundice, haemorrhaging and a variety of neurological symptoms.

REVIEW QUESTIONS

1.   What are the functions of vitamin A?

2.   What are the symptoms of vitamin A deficiency?

3.   How can vitamin A status be determined?

4.   How is vitamin D synthesized?

5.   Give the chemical features of vitamin D deficiency.

6.   Write a note on toxicity.

7.   Explain vitamin E as antioxidant.

8.   Give the importance of vitamin K.

CRITICAL THINKING QUESTIONS

1. Is there a danger of deficiency if a vitamin is missing in your diet for 1 week?

2. Are fat-soluble vitamins lost during cooking?

3. How does vitamin E act as an antioxidant?

8

WATER-SOLUBLE VITAMINS

Did You Know?

- Pork is an excellent source of thiamine along with sunflower seeds and dried beans.

- Enriched grains and white rice contain extra thiamine, niacin, riboflavin, and iron.

- Milk is a good source of riboflavin.

- A deficiency in a major B vitamin, such as riboflavin, suggests other B vitamins will also be deficient, since most are found in similar foods.

- Absorption of vitamin B_{12} in elderly patients is poor because cells in the stomach commonly decrease the synthesis of a factor that is vital for vitamin B_{12} absorption.

- The amino acid tryptophan is converted into niacin in the body. Niacin needs then are met by consuming both niacin-containing foods and protein in the diet.

- A niacin deficiency causes severe dermatitis and skin redness, especially where the sun strikes, as well as diarrhoea and dementia.

- Nicotinic acid in pharmacological (high) doses can lower serum cholesterol levels. At the same time, it often causes redness of the skin and itching.

- Alcoholism is a major cause of B vitamin deficiencies. Alcohol decreases the absorption of vitamin B_6, thiamine and folate.

- The term folate comes from the word foliage. Good sources of folate include green leafy vegetables and sprouts, organ meats and orange juice.

- Vitamin B_{12} is present only in animal foods. It is not found in plants unless its presence is due to fermentation or contamination of the product from bits of soil or insects.

- Vitamin C enhances iron absorption. Vitamin C converts Fe^{3+} into Fe^{2+}; the Fe^{2+} form of iron is absorbed more readily.

VITAMIN B$_1$ (THIAMINE)

Thiamine or B$_1$, and aneurine is widely known for its role in preventing the deficiency disease "beriberi". Beriberi occurred in epidemic proportions in Japan, China and South-East Asia in the 19th century. Osler (1893) characterized "wet Beriberi" as a condition with general oedema, shortness of breath and sensory disturbances with paralysis. The word beriberi means "I can't, I can't" and probably refers to the sufferer's inability to achieve neuro motor coordination. Takaki (1906) demonstrated that the disease in Japanese soldiers could be reduced by substituting wheat bread for part of polished rice.

Eijkman in 1858 produced polyneuritis in fowls by feeding them with a diet consisting of washed polished rice and showed that they recovered when given extracts of rice polishings.

Funk (1911) isolated anti-beriberi factor from rice polishings. Crystalline thiamine was isolated from rice bran by Janeen and Donath in 1926.

B$_1$ is called the antineuritic factor.

CHEMISTRY

Thiamine is soluble in water and easily destroyed by heat or oxidation especially in the presence of baking soda. The term thiamine indicates that it is a sulphur- and nitrogen-containing substance.

FUNCTIONS

1. Vitamin B$_1$ is a part of the coenzyme TPP made up of thiamine with 2 molecules of phosphates attached. It is

required for carbohydrate metabolism. The accumulation of the intermediary products of metabolism when TPP is absent is believed to cause typical symptoms.

There are three stages in the metabolism of carbohydrate during which the absence of thiamine as part of a coenzyme leads to slowing or complete block of chemical changes.

- as a part of TPP thiamine is necessary for the decarboxylation of pyruvic acid as it is prepared to enter the Kreb's cycle. When thiamine is lacking, pyruvic acid tends to accumulate.

- TPP too plays a similar role in the decarboxylation of α-ketoglutaric acid (which is an intermediary product of both fat and carbohydrate metabolism) to succinic acid.

- Activating transketolase which is an enzyme necessary for the metabolism of glucose for the production of ribose which is the carbohydrate needed for the synthesis of RNA and fatty acids.

2. Thiamine helps in maintaining nerves in normal condition. In thiamine deficiency, degeneration of myelin sheath of peripheral nerves and ganglion cells of brain and spinal cord occur. It also helps in the transmission of high-frequency impulses at the nerve synapse either through production or release of neurotransmitters such as acetylcholine or serotonin or by forming complexes with other neurotransmitters.

3. Thiamine is essential for growth and body maintenance and normal functioning of body processes in cooperation with other vitamins and minerals.

4. Thiamine also plays a role in the conversion of the amino acid tryptophan to niacin.

5. Thiamine has several indirect functions in the body because of its role in energy metabolism like

- maintenance of appetite

  ☉ maintenance of muscle tone

  ☉ maintenance of healthy mental attitude

It is often called the morale vitamin because of its ability to restore a healthy mental state.

METABOLISM

Thiamine ingested in the food is available in free form or bound as thiamine pyrophosphate (TPP) or a protein–phosphate complex. Bound forms are split in the digestive tract after which absorption takes place in the first part of the duodenum by an active process involving sodium-dependant ATPase. A second passive process of absorption operates at thiamine concentrations more than 1 μmol or oral intakes more than 5 mg/day. This dual mechanism has clinical implications because ethanol inhibits the active but not passive process. Following absorption, about 30 mg of thiamine is phosphorylated and stored as TPP in different organs like heart, brain, liver and skeletal muscle about 2.7, 1.2, 1.0 and 0.7 μg/g respectively. The conversion of thiamine to TPP requires ATP. TPP acts as a coenzyme for a number of enzyme systems. Thiamine circulates in free form in blood and the excess is excreted in the urine with small amounts of its metabolites primarily thiamine diphosphate and disulphide. Urinary thiamine excretion decreases rapidly in thiamine deficiency indicating a renal conservation mechanism.

Thiaminase is an enzyme present in uncooked clams, some fishes and shrimps that splits thiamine molecule and inactivates it. This causes no problem as cooking inactivates thiaminase. Tea and a few other foods contain thiamine antagonists. Persons consuming large amounts of tea have the risk of developing deficiency (8 cups/day or more).

TRANSPORT

Thiamine is carried by the portal blood to the liver. Most of thiamine in plasma is mainly bound to albumin. The transport of thiamine into erythrocytes seems to be a facilitated diffusion process, whereas its entry into other cells is an active process.

TISSUE DISTRIBUTION AND STORAGE

The total amount of thiamine in the normal adult is approximately 0.11 mmol (30 mg). High concentrations are found in the skeletal muscle, heart, liver, kidneys and brain. Because thiamine is not stored in large amounts in any tissue, a continuous supply of thiamine is necessary.

EXCRETION

Thiamine and its metabolite are mainly excreted in the urine. Very little thiamine is excreted in the bile.

EFFECTS OF DEFICIENCY

Thiamine deficiency is common among alcoholics. Thiamine deficiency may result from a low dietary intake when the diet is very low in calories. Deficiency of this nutrient may also result from a failure of absorption, usually caused by abnormality in GI tract, folacin deficiency, inability of tissues to accumulate adequate stores, failure to utilize available thiamine or increased requirement in a high carbohydrate diet or alcohol, both of which need B_1 to be metabolized.

Symptoms

1. Loss of appetite or anorexia, nausea, the lack of muscle tone in the wall of the GI tract results in decreased gastric motility, a distended colon and constipation.

2. Mental depression and confusion, mood changes, fear and disorderly thinking. They show irritability, fatigue and headache and decrease in socialization.

3. Neurological changes, nystagmus or involuntary rapid eye movement also known as Wernick's syndrome, is a result of

changes in the CNS. Severe disturbances of posture and equilibrium occur. Peripheral neuritis in which nerves that control extremities fail to function properly usually affect leg first resulting in calf muscle tenderness, loss of ankle and knee jerks, tingling or pins and needles in the feet and numbness. Neuromuscular coordination is affected resulting in a decreased mechanical efficiency and work output. Motor speed and eye hand coordination decreases.

In adults beriberi takes two forms.

- In wet or oedematous beriberi the victim suffers from swelling of limbs, starting at the feet and progressing upward throughout the body causing difficulty in walking. The accumulation of fluid in heart muscle leads to eventual heart failure and death. Also pulmonary congestion, venous distention, tachycardia and high blood pressure.

- In dry or wasting beriberi a gradual loss of body tissue occurs and the victim becomes thin and emaciated. In both forms, symptoms include numbness in legs, irritability, vague uneasiness, disorderly thinking and nausea, suggesting involvement of nervous system.

Infantile beriberi occurs in infants, 2–5 months of age. Onset is rapid and unless treated immediately condition often results in death. The baby develops symptoms like cyanosis or too much CO_2 in the blood, tachycardia, a very fast heart beat and a change from a loud piercing cry to a thin, weak almost inaudible one accompanied by vomiting and constipation. It occurs most often in breast-fed than in bottle-fed infants because the lactating mother's dietary intake of thiamine is often too low to produce milk with enough thiamine to protect her infant. This situation is complicated by a transfer of pyruvate aldehyde, a toxic product of carbohydrate metabolism that accumulates in thiamine deficiency, to mother's milk.

Wernicke's–Korsakoff's syndrome is an acute deficiency of thiamine.

Wernicke's syndrome includes ophthalmoplegia—paralysis of eye muscle and nystagmus—rapid eye movement and ataxia which is incoordinated gait.

Korasakoff's syndrome includes disturbed memory, inability to learn anything new, confusion, anxiety and fear. The symptoms appear after several weeks of eye and gait changes.

Neurological changes affect the occular muscles cerebellum and brainstem which may result in hypothermia and hypotension and necrosis of nerves and myelin sheath of particularly the mamillary bodies in the thalamus.

A genetic predisposition to Wernicke's disease is suggested by decreased transketolase activity.

SOURCES

Rich sources   Rice polishings, wheat germ, dried yeast.

Good sources   Whole cereals, nuts and legumes.

Fair sources   Meat, fish, egg. milk, vegetables, fruits.

REQUIREMENTS

An intake that leads to a small excretion is believed to represent minimal needs.

Since B_1 is a part of coenzyme needed in at least 3 places in the metabolism of carbohydrate, RDA of all age groups are based on caloric intake.

0.6 mg/1000 cal and an additional +0.2 mg during pregnancy and +0.5 mg during lactation.

Studies of older people show that their need for B_1 is higher. The need for B_1 increases with a increased consumption of alcohol because it is necessary for the metabolism of acetaldehyde, an intermediary product in alcohol metabolism. A degeneration in the intestinal wall of alcoholics results in decreased absorption.

The need for thiamine is inversely related to amount of fat in the diet. When fat calories, replace carbohydrate calories less thiamine will be required. The need for B_1 increases when some antibiotics like sulphonamides are given.

EVALUATION OF THIAMINE STATUS

The most sensitive test available for the determination of thiamine status measures the effect of TPP on the activity of RBC transketolase an enzyme which can function only if TPP is present as a coenzyme.

The load test is another method where people with low levels of saturation in the tissues will retain more and excrete less.

Factors Affecting Thiamine Status

Thiamine status depends on its bioavailablity in food products, ethanol consumption, presence of antithiamine factors (ATF) in the diet and folate and protein status.

Ethanol ingestion Thiamine deficiency in chronic alcoholics is caused by multiple factors, which include a low thiamine intake, impaired intestinal absorption, defective phosphorylation and an apotransketolase deficiency.

Antithiamine factors (ATF) Two types of ATF exist—thermolabile and thermostable. The thermolabile ATF include thiaminase I and II. Thiaminase I cleaves thiamine by an exchange reaction with an organic base or a sulphydryl compound via a nucleophilic displacement on the methylene group of the pyrimidine moiety of thiamine.

Thiaminase II is found in several microorganisms. These compounds are found in freshwater fish, shellfishes and sea fishes, ferns and in microorganisms.

Thermostable ATF have been found in ferns, tea, betel nut, some vegetables and in some animal tissues.

Some polyphenols also have antithiamine activity and oxygen. Quinones interact with SH form of thiamine to give thiamine disulphide. Further hydrolysis and oxidation yield inactive products. Ascorbic acid and other reducing agents prevent the formation of quinones and thiamine disulphide. Calcium and magnesium augment precipitation of thiamine by tannins making thiamine less bioavailable.

Folate and protein status   Thiamine is poorly absorbed in subjects with folate or protein deficiency.

LOSSES DURING PROCESSING AND PREPARATION

The extent to which foods lose thiamine in preparation is determined by the physical and chemical properties of the vitamin.

1. **Loss in solution**   Since thiamine is water-soluble, it will leach out of a product in proportion to the amount of water available, the extent to which it is agitated and the surface area of the food exposed to the water. Any method of preparation that minimizes the length of time a food is in contact with water and the surface area will decrease thiamine losses. As much as 18% of the thiamine in rice is reportedly lost in the oriental method of washing the rice several times before cooking. Modern marketing procedures which protect the food from contamination from the air, eliminate the necessity of preliminary washing of rice. In fact, most packages warn the housewife not to wash rice and to cook it in minimum water to reduce cooking losses. This is especially important when the rice is enriched by coating it with an enrichment mixture.

2. **Loss due to heat**   Thiamine is destroyed by heat. The higher the temperature and more prolonged the exposure to heat the greater the loss. Roasting pork at 162.9°C allows a retention of 75% to 100% of the original thiamine, whereas higher temperatures give lower yields. Destruction appears

no greater in cooking in electronic ovens than in conventional ones. Thiamine in food is less susceptible to heat destruction than in the free form. Also, a difference is found in rate of heat destruction between various foods. For example, thiamine in spinach, heart, and liver is more susceptible to heat destruction than that in peas, beans, pork and carrots.

3. **Loss due to oxidation** Cooking procedures that increase the amount of O_2 in contact with the food, especially under conditions of moist heat, speed up the destruction of thiamine. The use of rapidly boiling water in cooking vegetables is an example for this.

4. **Loss due to alkali** Destruction of thiamine is greatest in the presence of alkali. The addition of baking soda and alkali to cooking water, is sometimes suggested as a means of preserving the bright green colour of fresh vegetables. Its use for such purposes cannot be recommended because of the destructive effect it has on both thiamine and vitamin C.

5. **Loss due to processing** Thiamine content of pork is virtually destroyed by the irradiant procedures sometimes used in food preservation. The use of sulphite as preservative, especially in ground meat, leads to a reduction in thiamine activity.

VITAMIN B_2 (RIBOFLAVIN)

Riboflavin which has also been known as vitamin B_2, vitamin G and "the yellow enzyme" was recognized as a vitamin in 1917.

Although as early as 1897, a London chemist had observed in milk whey a water-soluble pigment with peculiar yellow-green fluorescence, it was not until 1932 that riboflavin was actually discovered by researchers in Germany. It was given the chemical group name flavins from the Latin word for yellow. Later because the vitamin was found also to contain a sugar named ribose, the name riboflavin was adopted.

CHEMICAL AND PHYSICAL NATURE

Riboflavin is a yellow-green fluorescent pigment that forms yellowish brown, needle-like crystals. It is water-soluble and relatively stable to heat but is easily destroyed by light and irradiation. It is stable in acid media and is not easily oxidized. However, it is sensitive to strong alkalis.

Riboflavin was chemically specified as 7, 8 dimethyl-10-(1´-D-ribityl) isoalloxazine.

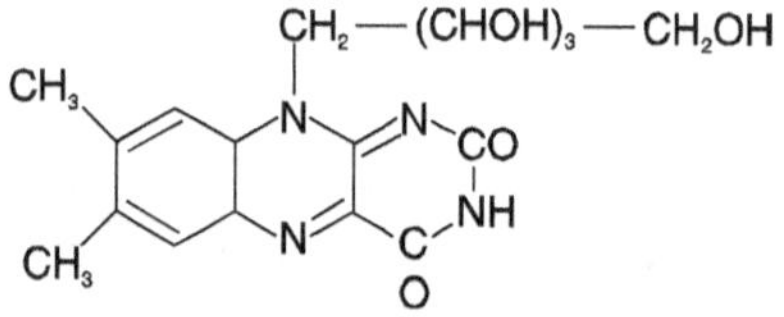

Riboflavin

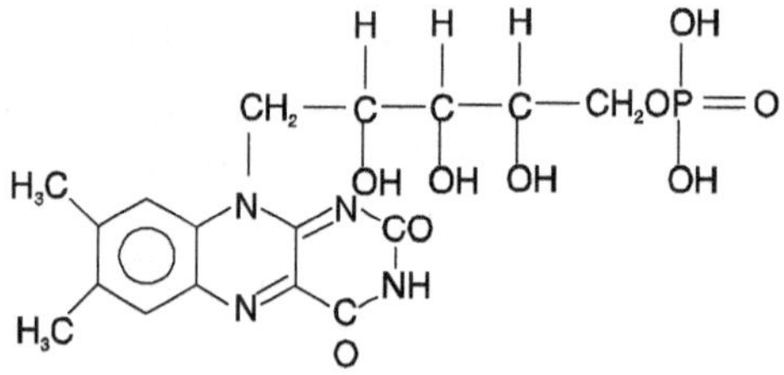

FMN

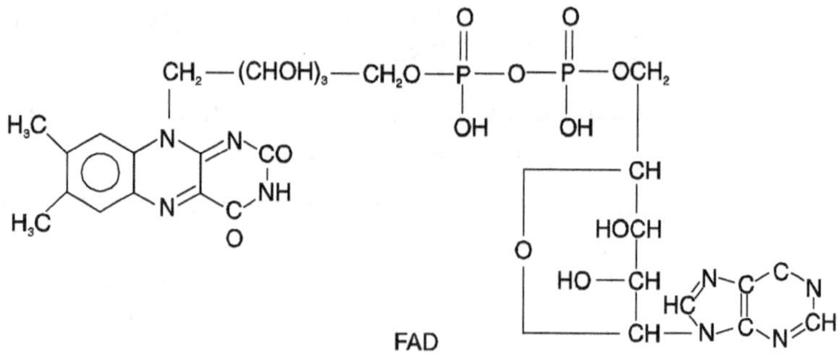

FAD

ABSORPTION, TRANSPORT, METABOLISM AND EXCRETION

Coenzyme forms of the vitamin (mainly FAD and less of FMN) are released from non-covalent attachment to proteins as a consequence of gastric acidification. Non-specific action of pyrophosphatase and phosphatase on the coenzyme forms occurs in the upper gut.

The vitamin is primarily absorbed in humans in the proximal small intestine by a saturable transport system that is rapid and proportional to dose. Bile salts appear to facilitate the uptake. Active transport at lower levels of intake may be Na-dependant and involves phosphorylation.

In humans, some of the riboflavin circulating in blood plasma is loosely associated with albumin, though significant amount of the vitamin complexes with IgG and several other immunoglobulins.

Metabolic conversion of riboflavin to coenzymes occur with the cellular cytoplasm of most tissues but particularly in the small intestine, liver, heart and kidney. The conversion is given below (Figure 8.1).

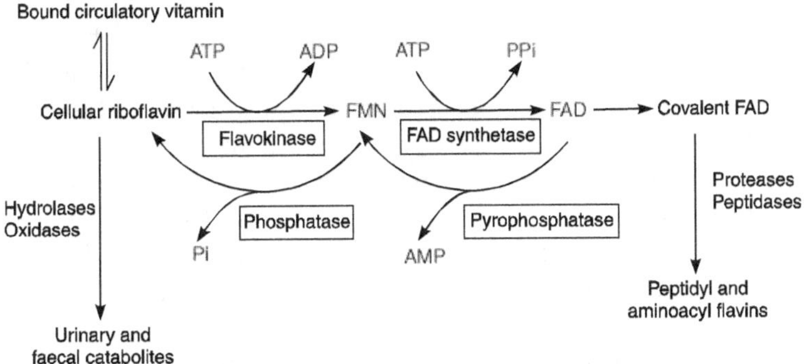

Figure 8.1   Conversion of riboflavin to coenzymes

Source   *Modern Nutrition in Health and Disease* edited by Shills, *et al.*

BIOCHEMICAL AND PHYSIOLOGICAL FUNCTIONS

Basic coenzyme role   The cell enzymes of which riboflavin is an important part are the flavoproteins. Riboflavin enzymes operate at vital reaction points in the process of cellular energy metabolism and in deamination, the key reaction that removes the nitrogen-containing amino group from certain amino acids. Thus riboflavin acts as a control agent in both energy production and tissue building.

Riboflavin is necessary for the conversion of tryptophan to the vitamin niacin.

Riboflavin also aids in the conversion of both vitamin B_6 and folacin to their coenzymes and in their storage in the body. Because these coenzymes are needed for DNA synthesis, riboflavin has an indirect effect on cell division and therefore on growth.

Riboflavin plays a role in the production of hormones in the adrenal cortex, in the formation of red blood cells in bone marrow, the synthesis of glycogen and also in fatty acid catabolism.

DEFICIENCY

The many manifestation of lack of riboflavin can be assessed clinically. In humans, an early form of ariboflavinosis is a condition known as cheilosis in which cracks appear at the corners of the mouth, and the lips become inflamed. There is a combination of symptoms, which centres on tissue inflammation and breakdown, and poor wound healing. Even minor injuries easily become aggravated and do not heal easily. The lips become swollen, cracking easily and characteristic cracks develop at the corners of the mouth, a condition called cheilosis. Cracks and irritation develop at nasal angles. The tongue becomes swollen and reddened, a condition called glossitis (Figure 8.2).

Extra blood vessels develop in the cornea—corneal vascularization—and the eyes burn, itch and tear. A scaly greasy skin condition—seborrhoeic dermatitis—may develop especially in skin folds.

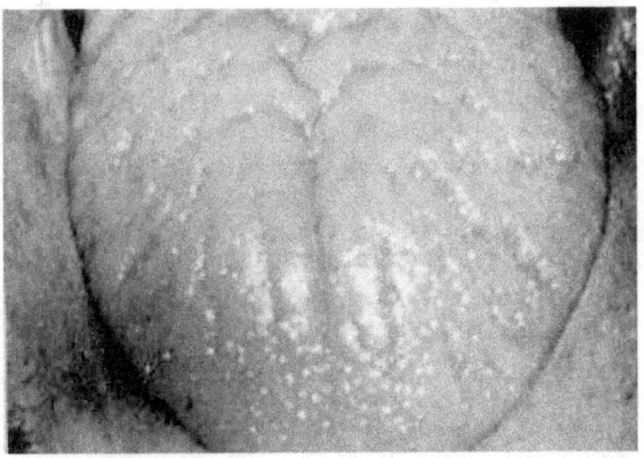

Figure 8.2   An inflammed tongue (glossitis) can signal a vitamin deficiency

Deficiency in Newborns

Since riboflavin is light-sensitive, newborn infants with hyperbilirubinaemia treated with phototherapy have shown signs of riboflavin deficiencies even when supplements were provided.

Treatment   A dose of 5 and 10 mg is adequate for curing oral and dermal lesions. Symptoms disappear in few days to few weeks.

FOOD SOURCES

The most important food source is milk. Lactoflavin the milk form of riboflavin is found in milk. Other good sources are organ meats such as liver, kidney and heart, whole or enriched grains or vegetables. Since riboflavin is water-soluble and destroyed by light, considerable loss can occur in open, excess-water cooking.

RIBOFLAVIN REQUIREMENT

Riboflavin requirement depends on the total energy needs, level of exercise, body size, metabolic rate and rate of growth. The general RDA standard for riboflavin is 0.6 mg/1000 kcal for all ages.

RISK GROUPS

People living in poverty, with gastrointestinal disease or with a chronic illness where appetite is poor and malabsorption exists, children, and pregnant and lactating mothers are more prone to deficiency.

Anorexia, intestinal malabsorption, chronic alcoholism and biliary atresia, all precipitate deficiency.

In experimental deficiency induced by a riboflavin antagonist, glossitis was followed by seborrhoeic dermatitis, peripheral neuropathy and hypoplastic anaemia.

Thyroxine deficiency may impair the conversion of riboflavin to FMN and diabetics are at risk because of high urinary excretion of the vitamin.

The drug chlorpromazine impairs FAD synthesis, and barbiturates induce microsomal oxidation and therefore riboflavin catabolism.

ASSESSMENT

Assessment involves measuring riboflavin in urine and red blood cells by fluorometric and microbiological techniques or determining the activity of the riboflavin-dependant enzyme, erythrocyte glutathione reductase (EGR). Urinary excretion reflects daily intake of the vitamin.

VITAMIN B_3 (NIACIN)

Niacin includes nicotinic acid and its amide nicotinamide, as well as any derivative that can be biologically converted to active compounds. It is also known as B_3.

Pellagra, the classic disease associated with niacin deficiency was first recognized in 1735 by the Spanish physician, Casals. Pellagra was rampant among the maize-eating population throughout the world. In the early 1900s the disease became an epidemic in the South-Eastern United States, and it was there where Goldberger established that pellagra was not an infectious disease but was caused by a dietary deficiency.

In 1945 the amino acid tryptophan was shown to replace niacin in improving the growth of rats fed with high-corn diet.

CHEMISTRY

Niacin is the generic description for the specific vitamin nicotinic acid as well as its natural derivative nicotinamide or niacinamide. Nicotinamide is incorporated within NAD or NADP.

Nicotinic acid
(niacin)

Nicotinamide acid
(niacinamide)

Pyridine nucleotide coenzymes

Structures of vitamins and the two coenzyme forms containing the nicotinamide moiety.

Nicotinic acid was first isolated by Funk in 1911 from rice polishings, but its significance was not known then. It is a pyridine β-carboxylic acid, with an empirical formula of $C_6H_5O_2N$. It is a white crystalline compound and sparingly soluble in water. It is stable to heat.

Nicotinamide is formed by the addition of an NH_2 radical to nicotinic acid. In animal tissues nicotinic acid exists as nicotinamide only. When taken orally, nicotinic acid is rapidly converted into nicotinamide.

Nicotinamide in food is not destroyed by cooking but can be lost if excessive water is used and drained off.

METABOLISM

NAD and NADP, the chief dietary forms of niacin, are hydrolysed by enzymes in the intestinal mucosa to yield nicotinamide as the major end product. Intestinal bacteria can convert nicotinamide to nicotinic acid. Both are absorbed by facilitated diffusion at lower concentrations and by passive diffusion at higher concentrations.

Niacin is rapidly removed from blood plasma by the tissues, especially the liver and the red cells. Although most tissue cells absorb niacin by passive diffusion, facilitated diffusion also takes place in some cells such as those of the kidney and the erythrocyte.

STORAGE AND EXCRETION

Once niacin enters the cell, it is converted to its coenzyme forms. In addition to NAD bound to enzymes, NAD not attached to an apoenzyme may also be present. This free NAD is sometimes designated as "storage" NAD.

In the liver any excess of free niacin that accumulates, is methylated to *n*-methyl nicotinamide by *N*-methyl transferase.

NMN is the major niacin metabolite excreted in the urine. The oxide and hydroxyl forms of niacin are also excreted in small amounts.

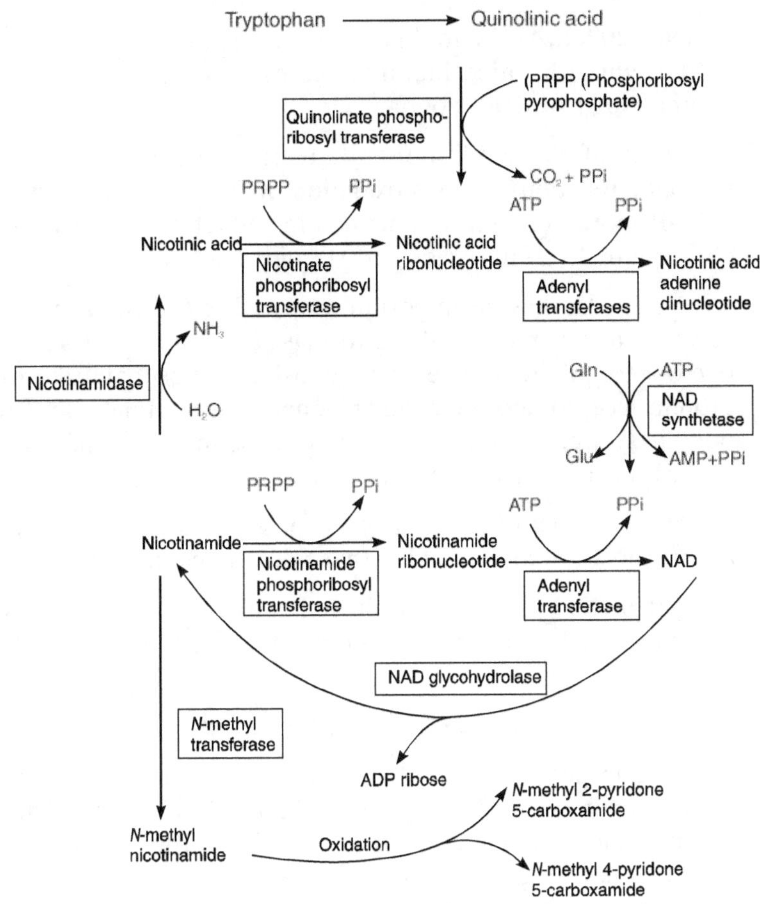

Figure 8.3　Pathways of niacin metabolism

Figure 8.3 shows the pathways of niacin metabolism. In the liver and kidney of mammals, quinolinic acid is converted to nicotinic acid ribonucleotide, from which the coenzymes NAD and NADP are synthesized.

FUNCTION

At least 200 enzymes are known to be dependant on NAD and NADP with the nicotinamide moiety acting as an electron acceptor or hydrogen donor.

Most of the NAD-dependent enzymes are involved in catabolic reactions such as oxidation of fuel molecules whereas NADP more commonly functions in reductive biosynthesis of compounds like fatty acids and steroids.

NAD also has an important non-redox function, in that it serves as a substrate for glycohydrolase catalysing the transfer of one or more ADPR moieties (adenosine diphosphate ribose) to protein acceptor molecules and regenerating nicotinamide. These polyribosylated proteins appear to function in DNA repair, DNA replication and cell differentiation.

Nicotinic acid is a component of glucose tolerance factor, an organochromium complex that potentiates insulin response.

Niacin is essential for the normal functioning of skin, intestinal tract and nervous system.

TRYPTOPHAN–NIACIN RELATIONSHIP

Tryptophan present in the dietary protein is converted to niacin in the human body. 60 mg of tryptophan will yield 1 mg of niacin or one niacin equivalent. Since protein is approximately 1% tryptophan, 60 mg of protein will provide 600 mg of tryptophan or 10 niacin equivalents.

$$\text{Niacin equivalent} = \text{mg of preformed niacin} + \frac{\text{mg of tryptophan}}{60}$$

The conversion of tryptophan to niacin requires at least three other vitamins thiamine, pyridoxine and riboflavin. The various steps involved in the process of conversion of tryptophan to nicotinic acid is illustrated through a flow chart.

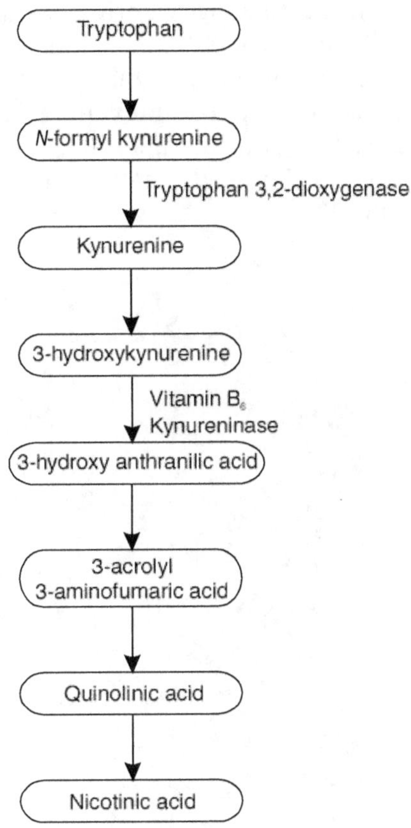

DEFICIENCY DISEASES

Pellagra

Pellagra is classically characterized by four D's.

1. Dermatitis
2. Diarrhoea
3. Dementia
4. Death

1. **Dermatitis** Pellagra derives its name from the skin manifestation, "pelle" = skin and "agra" = rough. Changes are marked on the skin exposed to the sun and friction. Erythema, thickening and pigmentation is followed by exfoliation, leaving a parchment like skin—Casal's necklace, dermatitis on the skin of the neck (Figure 8.4).

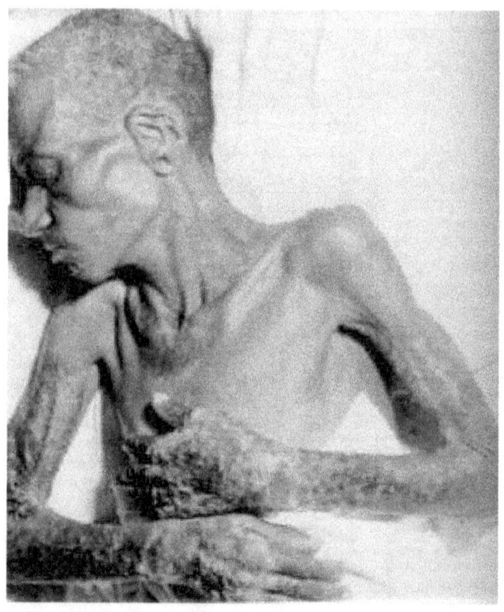

Figure 8.4 The dermatitis of pellagra

2. **Diarrhoea** Changes occur in the GI tract. The tongue appears raw, the papillae are lost and the mucous membrane of the mouth is inflamed. Achlorhydria is seen. The small intestine shows structural and absorptive defects.

3. **Dementia** Irritability, depression, poor concentration, loss of memory due to changes in the nervous system. This may be due to the formation of serotonin, a neurotransmitter. Anaemia may also occur. The lateral and posterior columns of spinal cord may be affected.

Treatment As nicotinic acid, when administered in large doses, causes flushing of skin and burning sensation, it is safer to administer nicotinamide for the treatment of pellagra. In severe cases of pellagra, nicotinamide can be administered intramuscularly in doses of 50 mg twice daily for 3–4 days, followed by oral dose of 100 mg thrice daily, or about 2–3 weeks till the disease is cured. A maintenance dose of 50 mg daily, may be continued for a month. The diet should contain large amounts of milk, liver, peanut, egg, meat and fish in addition to cereals.

Prevention In regions where pellagra is common, nicotinamide tablet (30 mg) should be distributed to the people. The staple food, usually corn or sorghum, should be fortified with nicotinic acid. The people should be encouraged to consume peanuts, meat and liver which are rich sources of niacinamide. Improved strains of maize and kaffir corn with higher quantities of tryptophan niacin and lower amounts of leucine should be introduced in these areas.

Hartnup's Disease (Hereditary pellagra)

It is a hereditary disease characterized by intermittent attacks of rough, scaly reddened skin on moderate exposure to sunlight. More severe exposure produces severe rash.

The primary defect is due to amino acid transport and absorption. Amino acids such as lysine, tryptophan, phenylalanine and arginine are not absorbed and pass on into the colon where typtophan is acted upon by the normal intestinal flora to form indole compounds which are absorbed and excreted in the urine.

Other Uses of Nicotinic Acid

1. *Vascular disease* Nicotinic acid (only) is a vasodilator and hence used in CVD.

2. *Pain of coronary thrombosis* Nicotinic acid (50 mg given subcutaneously) helps in relief of pain and shock.

3. *Diarrhoea* Nicotinic acid is useful in pancreatic diarrhoea and prevents cAMP accumulation in gut cells during to cholera. This causes loss of fluids and electrolytes.

LEUCINE CONTENT OF MAIZE AND SORGHUM

The extensive studies of Gopalan and co-workers in India have shown that high leucine content of maize is one of the major factors contributing to the development of pellagra on maize diets. They have shown that leucine interferes with the conversion of tryptophan to niacin and also with the conversion of niacin to NAD and NADP in the tissues. They also found that poor kaffir corn (*Sorghum vulgare*) diet, which is rich in leucine like poor maize diet, causes pellagra in human beings and black tongue in dogs. They also found that a synthetic diet containing casein with added leucine can produce black tongue in dogs.

It may be concluded that nutritional defects in maize giving rise to pellagra are

1. low nicotinic acid content

2. low tryptophan

3. high leucine contents of maize proteins

It is also possible that poor maize diets are deficient in riboflavin, folic acid and vitamin B_{12} in addition to niacin which may be responsible for the anaemia, spinal cord changes and scrotal dermatitis observed in subjects suffering from pellagra.

In wheat, niacin occurs as niacytin, which is usually not available at all and is excreted in the urine.

SOURCES

Rich sources Dried yeast, liver, rice polishings, peanut.

Good sources Whole cereals, legumes, meat, fish.

Fair sources Milled cereals, maize, roots and tubers, other vegetables, milk, egg.

REQUIREMENTS

Since, tryptophan present in dietary proteins is converted into niacin in the tissues, a consideration to niacin requirements should also include niacin provided by tryptophan in addition to free niacin present in the diet. Horwitt introduced the term "Niacin equivalent", which is defined as 1 mg of niacin or 60 mg of tryptophan.

As niacin is crucial in releasing the energy from nutrients, RDA for niacin is based on the caloric intake. The minimum amount of niacin needed to prevent pellagra has been established at 4.4 niacin equivalents (NE) per 1000 kcal. A 50% margin of safety to take into account individual variations has been added to RDA, resulting in recommended allowances of 6.6 NE/1000 kcal. Regardless of the caloric intake, a daily minimum intake of 13 NE is recommended by FAO and WHO.

MEGAVITAMIN THERAPY

Large doses of niacin have been used in attempts to reduce blood cholesterol levels, which is a known risk factor in coronary heart diseases, and also to treat schizophrenia.

Under strict medical supervision, doses of 1 to 2 g of niacin (not nicotinamide) 3 times a day may result in lowered blood cholesterol levels and may be beneficial in protecting against recurrent non-fatal heart attacks.

In treating psychiatric problems, not only niacin but also other vitamins have been used in large amounts, in what is known as orthomolecular therapy. The use of large doses of niacin in the treatment of schizophrenia is controversial, and there is little scientific evidence to support its effectiveness.

ANTAGONIST

3-pyridine sulphonic acid acts as an antivitamin to niacin.

VITAMIN B_6 (PYRIDOXINE)

Pyridoxine refers to a group of naturally occurring pyridine derivatives that are metabolically and functionally interrelated.

Gyorgy identified and separated the heat-labile vitamin B_6 that cured a scaly dermatitis in rats fed with purified diets. The structure and synthesis of vitamin B_6 or pyridoxine was established in 1939 and additional forms of pyridoxine and its active form as pyridoxal phosphate were defined during World War II.

BIOCHEMISTRY

Vitamin B_6 occurs naturally in three forms namely pyridoxine, pyridoxal and pyridoxamine.

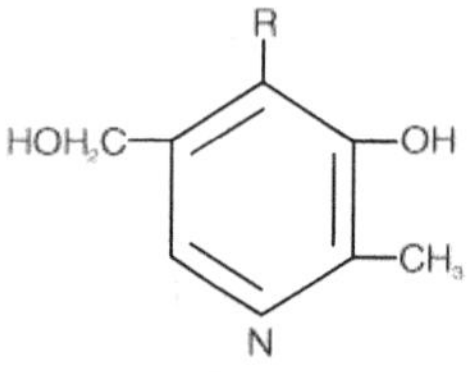

R / HOH$_2$C— —OH —CH$_3$ / N

If R = CH$_2$OH, pyridoxine (pyridoxol)
If R = CHO, pyridoxal
If R = CH$_2$NH$_2$, pyridoxamine

All three natural forms of vitamin B_6 undergo phosphorylation in the 5-position and oxidation to the active coenzyme pyridoxal phosphate.

The free forms of vitamin B_6 are considered relatively labile with the degree of lability influenced by pH. All three forms are relatively heat-stable in an acid medium but they are heat-labile under alkaline conditions.

METABOLISM AND ABSORPTION

The predominantly phosphorylated forms of dietary B_6 are hydrolysed by intestinal alkaline phosphatases before being absorbed in proportion to the luminal concentration.

The three primary forms of vitamin B_6 are absorbed to a major extent by a nonsaturable passive process mainly in the jejunum.

Rephosphorylation occurs in the liver, kidney and brain before conversion to the common form of the vitamin to pyridoxal phosphate by an oxidase. Renal aldehyde oxidase converts unbound PLP to pyridoxic acid, which is the principal metabolite excreted by the kidney.

BIOAVAILABILITY

Generally the availability of vitamin B_6 is greater than 75% in most foods studied. In some of the studies, there appears to be an inverse relationship between the amount of pyridoxine glucoside in the food and the bioavailability of vitamin B_6.

FUNCTIONS

Vitamin B_6 in the form of pyridoxal phosphate functions as a coenzyme for many reactions.

1. **Gluconeogenesis**   Pyridoxal phosphate is involved in gluconeogenesis via its role in transamination reactions and in the action of glycogen phosphorylase. Glycogen phsophorylase activity in liver and muscle is reduced in vitamin B_6-deficient rats.

2. **Niacin**   The direct conversion of tryptophan to niacin involves a pyridoxal phosphate requiring the enzyme, kynureninase. Hence a low level of pyridoxine affects niacin synthesis from tryptophan.

3. **Lipid metabolism**   It is necessary for the conversion of linoleic acid to arachidonic acid and for the synthesis and turnover of cholesterol.

4. **Erythrocyte metabolism and function**   In the erythrocyte pyridoxal phosphate serves as a coenzyme for transaminases. Both pyridoxal phosphate and pyridoxol bind to haemoglobin. Pyridoxol binds to the α-chain of haemoglobin and increases oxygen-binding affinity, whereas pyridoxal phosphate binds tightly to the β-chain and lowers oxygen-binding affinity. A severe chronic deficiency of vitamin B_6 can lead to hypochromic microcytic anaemia.

It is also necessary for the synthesis of the iron-containing protein of haemoglobin molecule.

EFFECT ON NERVOUS SYSTEM

Pyridoxal phosphate is a coenzyme for enzymatic reactions that led to the synthesis of several neurotransmitters, including serotonin from tryptophan, taurine, dopamine, norepinephrine, histamine and γ-amino butyric acid. Neurological abnormalities in human infants and animals deficient in vitamin B_6 have also been seen. Abnormal EEG tracings were also observed with convulsions.

EFFECT ON HORMONE MODULATION

Pyridoxal phosphate binds to steroid receptors and inhibits the binding of the steroid receptor to DNA. This results in a decreased action of the steroid. Reactions between pyridoxal phosphate and receptors for estrogen, androgen, progesterone and glucocorticoids suggest that the vitamin B_6 status of an individual may have significance in endocrine-mediated diseases.

VITAMIN B_6 DEFICIENCY

The clinical signs of accompanying vitamin B_6 deficiency usually occur in the latter stages of deficiency and are often seen with other water-soluble vitamins.

Some of the clinical signs of vitamin B_6 deficiency seen primarily in infants are abnormal electroencephalogram pattern and convulsion and those seen primarily in adults are stomatitis, cheilosis, glossitis, irritability, depression and confusion.

Aetiology

The factors leading to vitamin B_6 deficiency are as follows:

1. Inadequate dietary intake

2. Impaired delivery of vitamin B_6 due to

    i. Defective intestinal absorption

    ii. Impaired oxidation of pyridoxine

    iii. Impaired phosphorylation to form active coenzyme.

3. Excessive loss of vitamin B_6 through

    i. Kidneys

    ii. Oxidation

    iii. Inactivation by drugs

4. Relative deficiency (Primary intake inadequate relative to demand) due to

    i. increased metabolic activity (during pregnancy, fever, etc.)

    ii. Increased protein intake

5. Metabolic defects that alter use.

The widespread distribution of vitamin B_6 in foods prevents its deficiency to a great extent. However, several possible disorders have been observed.

1. Severe deficiency leads to hypochromic microcytic anaemia. However, anemia can be treated if vitamin B_6 is administered. Symptoms are fatigue, giddiness, nausea, insomnia, anorexia, palpitation, etc.

2. Deficiency leads to CNS disturbances. Infants show signs of irritability and convulsions which disappear when vitamin B_6 is administered.

3. Deficiency may lead to nausea and vomiting during pregnancy.

4. Other symptoms include weakness, insomnia, nervousness and difficulty in walking.

5. In pyridoxine deficiency, there is a decrease in the amount of urinary citrate which aids in the solubility of oxalates. This may cause the formation of urinary calculi. The lack of transaminases necessary to convert oxalic acid to glycine may also be responsible.

DRUG–VITAMIN B$_6$ INTERACTION AND TOXICITY

In many cases the drugs either react with pyridoxal phosphate or interfere with vitamin B$_6$ metabolism.

Adults on intake of 200 mg/day for one month showed signs of vitamin B$_6$ dependency. However vitamin B$_6$ has now been considered non-toxic.

SOURCES

Rich sources White meats, liver, whole grain, cereals and egg yolk.

Good sources Bananas, potatoes and avocados.

Fair sources Milk, cheese and citrus fruits.

REQUIREMENTS

The ICMR-recommended allowances of pyridoxine is given in the following table.

Group	mg/day pyridoxine
Man and Women	2.0
Pregnancy and lactation	2.5
Infants	
0–6 months	0.1
6–12 months	0.4
Children	
1–6 years	0.9
7–9 years	1.6
Boys and girls	
10–12 years	1.6
Adolescents	2.0

ASSESMENT

Assessment of vitamin B_6 status can be grouped into three categories—direct, indirect and dietary intake (Table 8.1).

Table 8.1 Indices of vitamin B_6 status and suggested minimal values for adequate status

Index	Adequate status
Direct	
Plasma pyridoxal phosphate	> 30 nmol/L
Plasma total vitamin B_6	> 40 nmol/L
Urinary 4-pyridoxic acid	> 3.0 μmol/L
Urinary total vitamin B_6	> 0.5 μmol/L
Indirect	
Erythrocyte alanine transaminase index	< 1.25
Diet intake	
Vitamin B_6 intake weekly average	> 1.2–1.5 mg
Vitamin B_6 protein ratio (mg/g)	> 0.016

Source *Modern Nutrition in Health and Disease* edited by Shills, *et al.*

ANTAGONISTS FOR VITAMIN B_6

Isonicotinic acid hydrazine (Isoniazid) used for treating tuberculosis inactivates pyridoxal phosphate by forming a hydrazone with it. So patients receiving 300 mg or more isoniazid require a daily dosage of 10 mg to meet this requirement.

Also oral contraceptic pills containing oestrogen increase the requirements of pyridoxine.

VITAMIN B_5 (PANTOTHENIC ACID)

Pantothenic acid, identified as vitamin B_5, was so named to designate its widespread occurrence in foods (*"pantos"* means everywhere). Its central role as part of coenzyme A and its functions in the metabolism of carbohydrate, fat and protein has been recognized.

It was isolated in 1939 and chemically identified in 1940. It was found to be effective in the cure or prevention of dermatitis in chicks.

CHEMISTRY

Pantothenic acid is a dimethyl derivative of butyric acid linked to β-alanine. The vitamin is linked through phosphate to form 4´ phosphopantetheine and CoA.

Pantothenic acid is a water-soluble vitamin that is stable in moist heat and in neutral solution, but is relatively unstable in dry heat, acid or alkali.

Chemically, pantothenic acid is a relatively simple compound containing the non-essential amino acids alanine and pantoic acid.

FUNCTIONS

1. Pantothenic acid being a part of coenzyme A, participates in the release of energy from carbohydrates, protein and fat.

2. It is necessary for the synthesis of fat.

3. CoA is involved as a source or acceptor for the acetate groups.

4. It helps in the formation of acetylcholine, a neurotransmitter.

5. It is necessary for the following.

    - formation of porphyrin.

    - stimulation of antibody responses.

    - the synthesis of cholesterol.

    - synthesis of hormones produced by adrenal glands.

    - synthesis of leucine, arginine and methionine.

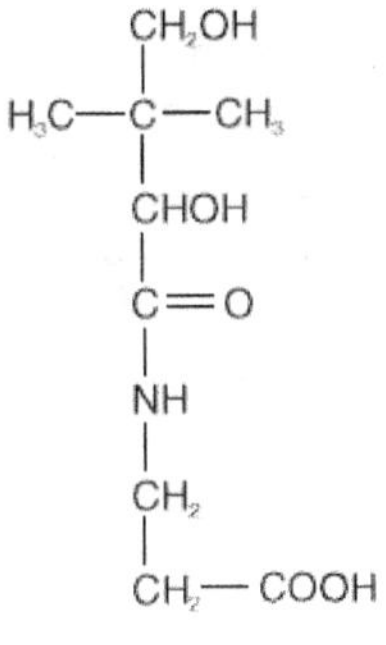

Pantothenic acid

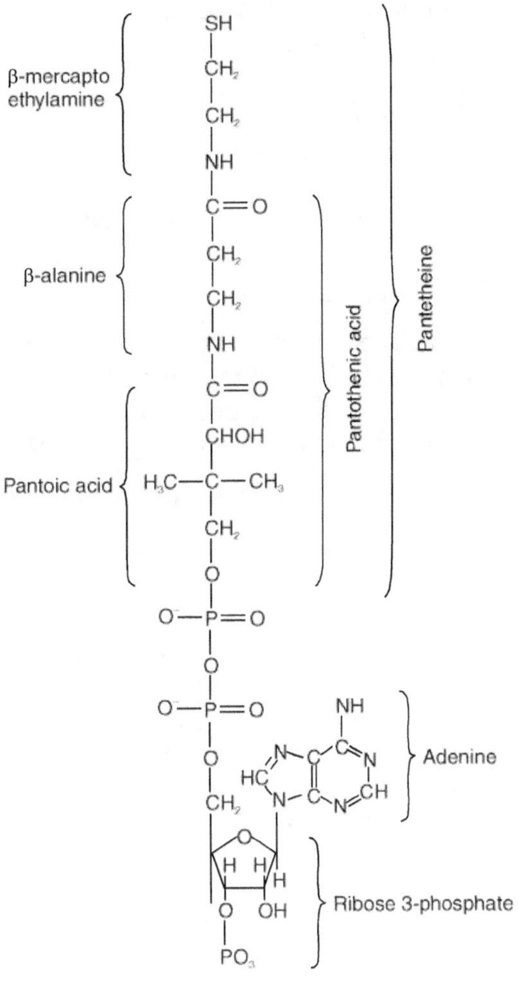

Coenzyme A

ABSORPTION, METABOLISM AND EXCRETION

Pantothenic acid is ingested as the part of CoA, which is hydrolysed by intestinal phosphatases to yield 4-phosphopantetheine and pantothenic acid, the absorbable form of the vitamin. CoA is resynthesized in liver cells.

Pantothenic acid is excreted in the urine, manually as a metabolic product of CoA.

FOOD SOURCES

Rich sources   Liver, kidney, egg yolk and milk.

Fair sources   Meat, cheese, legumes and vegetables.

REQUIREMENTS

Defieciency rarely affects the adrenal cortex, the nervous system, skin and hair. In rats, deficiency leads to impaired antibody formation, inflammation of the respiratory tract, anaemia, loss of hair, pigmentation and reproductive failure.

The RDA's safe and estimated range for adults is 4 to 6 mg.

VITAMIN H (BIOTIN)

In 1927, Boar made an important observation that when raw egg was incorporated as the main source of protein in the diet of rats, they developed deficiency symptoms characterized by dermatitis, loss of hair and muscle incoordination. All these symptoms were cured by yeast, live and egg yolk. The factor was called anti-egg white factor.

Later Gyorgy (1931) gave the name vitamin H to this factor. Vitamin H was isolated from egg yolk by Kogel and Tannis in 1936. Vitamin H was also isolated from milk by Merville in 1942 and this vitamin H is now called biotin.

Biotinidase is an enzyme which releases biotin from a protein and functions as a biotin carrier protein in human serum.

PROPERTIES

Biotin consists of an imidazole ring. Biotin is stable to heat and labile to strong acids and alkali. It is readily soluble in hot water, sparingly soluble in cold water and alcohol and insoluble in fat solvents.

Biotin combines with avidin, a glycoprotein found in egg white, to form a stable complex which is not broken down by proteolytic digestion. One molecule of avidin can combine with three molecules of biotin. 1 mg of avidin can combine with 138 mg of biotin. When biotin is so combined, it is non-absorbable and nutritionally unavailable, but avidin present in egg can be denatured by prolonged heating (steaming at 100°C) for 60 minutes.

ABSORPTION

Biotin is absorbed readily from the small intestines through the portal vein into the general circulation. Biotin ingested in excess of requirements is not stored but excreted in the urine.

FUNCTIONS

Biochemical

1. *Carboxylation reaction*   Biotin coenzyme is essential for the introduction of carboxyl group in some compounds. Propionyl-CoA carboxylase is involved in the conversion of propionic acid to succinic acid. β-methyl crotonyl-CoA

carboxylase catalyses the conversion of β-methyl crotonyl-CoA to β-methyl glutaconyl-CoA in the presence of biotin. Biotin acts as the coenzyme in several carboxylases used in metabolism.

2. *Transcarboxylase reaction* It catalyses this reaction as follows:

$$\text{Pyruvic acid} + \text{methyl malonyl-CoA} \xrightarrow{\text{Biotin}}$$
$$\text{oxaloacetic acid} + \text{propionyl-CoA.}$$

2. It is involved in the initial steps in the synthesis of some fatty acids.

3. It takes part in conversion reactions involved in the synthesis of some amino acids.

Physiological Functions

1. It is required for normal skin. In biotin deficiency, the skin shows extensive hyperkeratosis some parakeratosis, acanthosis and oedema.

2. Biotin is required for normal gestation and for lactation in rats.

3. Biotin deficiency causes paralysis of the hind legs in experimental animals.

4. Cell growth depends on some biotin-mediated activity.

5. It enhances protein synthesis, DNA synthesis and cell growth and also synthesis of nicotinic acid, purines, prostaglandins, pancreatic amylase and antibody formation.

6. The stimulation of testicular protein synthesis by testosterone also requires normal biotin status.

7. Biotin enhances the amount of translatable mRNA coding for glucokinase.

DEFICIENCY IN HUMANS

In 1942, Syndenstricken and his colleagues produced biotin deficiency by feeding four volunteers, a diet very poor in all the vitamins of the B group. This diet was composed of egg white which supplies 30% of the energy in the diet. Other known compounds of B-complex except biotin were added in synthetic form. After 10 weeks, the subjects were fatigued, depressed, sleepy with nausea and loss of appetite, muscular pain, hyperasthesia, and parasthesia developed. The tongue becomes pale with loss of papillae, the skin becomes dry, cracked with fine scaly desquamation, anaemia and hyperchlesterolaemia developed. All these were relieved by a concentrated preparation of 150–300 mg biotin daily.

Biotin deficiency has been reported in an adult who was known to consume 4–12 raw eggs and 1–4 quarts of wine; he suffered from severe dermatitis which responded to injections of methyl esters of biotin.

A second case of deficiency was noted in a boy with poliomyelitis who was receiving up to 6 raw eggs daily by way of a gastric tube. He developed scaly dermatitis and loss of hair.

Infants   This disease in young infants was reported in 1948 by Brown Deficiency often occurs in breast-fed infants along with persistent diarrhoea. Human milk contains a low quantity of 0.7 mg biotin in 100 ml of milk. The low level of biotin content of human milk along with poor absorption of biotin due to diarrhoea appears to cause biotin deficiency.

TOXICITY

Toxicity due to excess consumption of biotin is not known, though excess is excreted and not stored.

SOURCES

Biotin is found in both animal and vegetable foods.

Dried yeast, rice, wheat germ, liver, peanut, and soyabeans are rich sources.

Whole cereal, legumes, meat, fish, and poultry are good sources.

Milled cereals, vegetables, fruits and human milk are fair sources.

REQUIREMENTS

Not been established though a daily requirement of 150–300 mg is adequate, also aided by the synthesis of natural intestinal flora.

ASSESSMENT

Biotin can be measured in whole blood or urine by microbiological assay. Normal individuals have whole blood levels of 0.22–0.75 mg/ml with urinary excretion of 6–50 µg/day.

VITAMIN B_{12} (CYANOCOBALAMIN)

In 1926 Minion and Murphy discovered that pernicious anaemia was cured when patients suffering with it were fed with large amounts of raw liver.

Later Castle discovered that both an extrinsic factor (present in food especially liver) and an intrinsic factor (present in normal gastric secretion) are necessary for treating pernicious anaemia. The extrinsic factor present in liver was later named as B_{12}.

CHEMISTRY

It contains the mineral cobalt which is 4% of its weight. The cobalt was present at the centre of a large, complex molecule known as corrinoid, which resembles haemoglobin or chlorophyll. With this discovery vitamin B_{12} was named cobalamin. One form of the vitamin, cyanocobalamin was bound to have a cyanide group closely found in the molecule along with cobalt. Other forms of the vitamin hydroxycobalamin or methyl cobalamin are found in dairy products. In body tissues, the hydroxyl and cyanide groups of the vitamin get replaced by adenosine and forms 5-deoxy adenosyl cobalamin which functions as a coenzyme. Since all these compounds contain cobalt, they still have vitamin B_{12} activity.

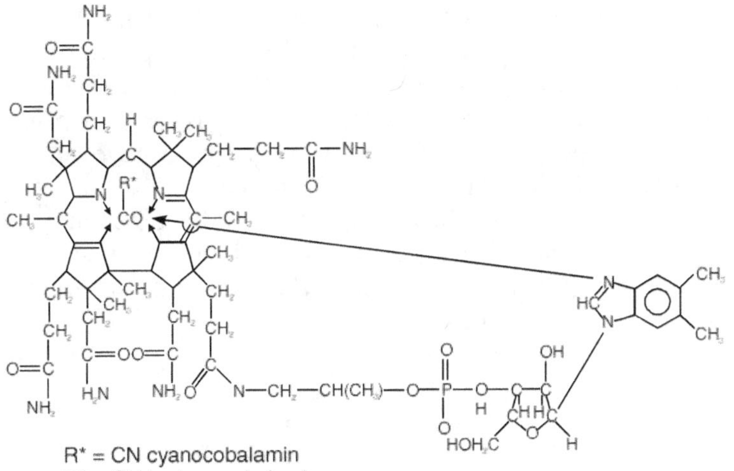

$R^* = CN$ cyanocobalamin
$R^* = OH$ hydroxocobalamin
$R^* = H_2O$ aquocobalamin
$R^* = NO_2$ nitrocobalamin
$R^* = CH_3$ methyl cobalamin

ABSORPTION AND STORAGE

Vitamin B_{12} is separated from its protein complex (polypeptide linkage) by gastric acid and enzymes. Vitamin B_{12} has a unique mechanism for its absorption. Even though the daily requirement is only a few micrograms, its absorption from the intestines requires a factor called **intrinsic factor** (IF) secreted by the stomach. IF binds vitamin B_{12} and the IF-B_{12} complex is absorbed. Dietary vitamin B_{12} and small oral doses of the vitamin are absorbed by the IF mechanism. Large oral doses are absorbed by diffusion. Normal subjects have serum levels of 20–90 m µg/100 ml with a total binding capacity of 50–110 m µg/100 ml.

The main storage organ of vitamin B_{12} is the liver. The amount in excess of 50 mg is rapidly excreted in urine.

Intrinsic Factor

Normal absorption of vitamin B_{12} requires that it be bound to a glycoprotein present in gastric juice called "intrinsic factor". When so bound, about 70% of dietary vitamin B_{12} is absorbed. When unbound, less than 2% is absorbed. Intrinsic factor is a thermolabile glycoprotein; it is sensitive to peptic digestion. Although most nutrients are absorbed mainly in the upper segments of the small intestines, vitamin B_{12} is absorbed principally from the lowest level of the ileum. The absorptive cells of the ileum are highly specific for this function. Upon reaching the ileum, the intrinsic factor-B_{12} complex becomes attached to specific receptor sites on the brush border of the mucosa cells, a process requiring a pH greater than 5.7 and the presence of cations, especially calcium.

TRANSPORT

Most of the B_{12} found in circulation is bound to one of two proteins in plasma. These are designated as transcobolamin I (TC I) and transcobalamin II (TC II). Transport of vitamin B_{12} is accomplished chiefly by TC II.

FUNCTIONS

Physiological Functions

1. B_{12} promotes maturation of the erythroid cells.

2. Vitamin B_{12} acts on other marrow elements as shown by the increase in white cells count and blood platelet count.

3. Vitamin B_{12} stimulates appetite and the general health of a person.

4. It cures the neurological symptoms of pernicious anaemia.

5. In the bone narrow where erythroblasts, the forerunners of RBC, are formed, B_{12} coenzymes provide the methyl groups for the synthesis of DNA. If DNA is not produced, the cells cannot divide, instead they continue to produce RNA and to synthesize protein, increasing in size. These large RBC or macrocytes which are characteristic of pernicious anaemic patients differ from the smaller mature RBC formed when cobalamin is available. The role of B_{12} in nucleic acid synthesis is important in all body cells.

Maintenance of Nervous Tissue

For the proper maintenance of the myelin sheath of nerve fibres vitamin B_{12} is necessary as B_{12}-dependent enzyme is necessary for the fatty acid synthesis for the lipoprotein complex of the myelin sheath. Any deficiency of vitamin B_{12} leads to impaired metabolism of fats and carbohydrate which are the major suppliers of energy to the nervous tissue. This in turn will lead to demyelination of the nerve fibre. Myelin sheath is a lipoprotein complex and B_{12} deficiency leads to a damaged myelin.

1. B_{12} is necessary for the synthesis of the amino acid methionine from homocysteine.

2. It facilitates the formation of the folate coenzyme needed for nucleic acid synthesis. B_{12} actually helps in the release of methyl folate reserve in the liver and blood serum. So it can be used as a coenzyme.

3. B_{12} is essential for the formation of single carbon units, which are then converted to many important compounds in the body (folacin aids in the transfer of these single carbon units).

4. OH cobalamin is a form of B_{12} which has an affinity for cyanide and this appears to be a detoxifying mechanism in people exposed to repeated small amounts of cyanide in food or tobacco smoking.

5. Vitamin B_{12} is essential as a component of animal protein factor (APF) which stimulated growth in animals. Penicillin stimulates growth in animals because it destroys organisms that destroy B_{12}.

    The use of B_{12} as a growth factor for children has been suggested.

6. B_{12} is involved in the both fat and carbohydrate metabolism as it is required for hydrogen transfer whereby methyl malonate is converted to succinate.

7. It is involved in making available more of the lipotropic substances choline and betaine and in protein synthesis through its role in the synthesis of the amino acid methionine.

8. B_{12} is involved in maintenance of sulphydryl (SH) groups in the reduced form necessary for function of many SH-activated enzyme systems. Vitamin B_{12} deficiency is characterized by a decrease in reduced glutathione (GSH) to oxidized glutathione (GSSG) of erythrocyte and liver.

9. B_{12} is essential for the release of folacin from the methyl folate reserves in the liver.

10. B_{12} also appears to be essential for the synthesis of the protein osteocalcin, needed for the formation of new bone cells.

CLINICAL DEFICIENCY

Vitamin B_{12} deficiency occurs as a result of acquired gastrointestinal disease and rarely as a result of inadequate diet or congenital disorders of absorption and transport.

Inadequate Diet

Because dietary vitamin B_{12} is exclusively of animal origin, strict vegetarians are at risk for vitamin B_{12} deficiency after many years on their diet. However deficiency is rare because many vegetarians in the developed world ingest multivitamin supplements containing vitamin B_{12} and because many vegetarians in the less developed world have enough bacterial contamination of the diet to obtain bacterially synthesized vitamin B_{12}.

Gastric Abnormality

Pernicious anaemia (PA) with gastric atrophy and deficient IF secretion usually occurs after age 40. Gastric atrophy is associated with destruction of the gastric acid, even in response to stimulation.

Small Intestinal Abnormality

Vitamin B_{12} deficiency is one consequence of the bacterial stasis syndrome which occurs in individuals with small intestinal diverticulosis, strictures or fistulas and in scleroderma with abnormal intestinal motility. In these syndromes, contaminating bacteria are presumed to take up and utilize vitamin B_{12} as it passes through the intestinal lumen.

Ileal Abnormality

Because of the exclusive presence of the vitamin B_{12}–IF receptor on the iteal surface, vitamin B_{12} malabsorption occurs when the ileum is diseased or surgically resected. Vitamin B_{12} malabsorption also occurs in tropical sprue but rare in coeliac disease, because the former mainly affects the ileum and the latter the jejunum.

Drug Interaction

Several drugs, e.g. ethanol, and *p*-amino salicylic acid produce vitamin B_{12} deficiency by interacting with the ileal vitamin B_{12}–IF receptor.

CLINICAL FEATURES AND TREATMENT

Vitamin B_{12} deficiency can result in abnormalities in bone narrow, small intestine and nervous system. The deficiency causes weakness, glossitis and diarrhoea.

1. Pernicious anaemia is a disease of genetic origin in which intrinsic factor is not produced and vitamin B_{12} is not absorbed. The bone marrow is unable to produce mature RBCs and releases fewer large cells (macrocytes) into the circulation so the capacity to carry haemoglobin is reduced. Characteristic symptoms include pallor, anorexia, decreased weight, increased bleeding, abdominal discomfort and mental depression.

2. Megaloblastic anaemia from B_{12} deficiency also occurs following surgical removal of the part of the stomach which produces the intrinsic factor or part of ileum where absorption sites are located. This can be prevented by B_{12} injections after surgery.

3. Lack of protein results in transcobalamin II which binds cobalt.

4. Worm infestation decreases absorption.

5. B_{12} is found in foods of animal origin, hence vegetarians need supplements.

6. Absorption of vitamin B_{12} decreases with increasing age. one-third of people over 60 do not secrete gastric acid so they cannot absorb vitamin B_{12} from food because they are unable to split it from the protein complex. They also secrete less intrinsic factor.

SOURCES

B_{12} is not synthesized in plants and hence is found only in foods of animal origin. Microorganisms in the GI tract of humans can also synthesize the vitamin, but the site of synthesis is far too down in the colon to permit absorption. Thus humans must depend on animal food for B_{12}.

Good sources   Liver, meat, fish, egg.

Fair sources   Milk and milk products.

REQUIREMENTS

ICMR-recommended allowances of vitamin B_{12} is given in the following table.

Group	Vitamin B_{12} μg/day
Man, woman, pregnant woman	1.0
Lactating mother	1.5
Infants	0.2
Children and adolescents	0.2–1.0

ASSESSMENT

Assessment is estimation of serum B_{12} which falls below 200 pg/ml in B_{12} deficiency.

Functional deficiency of vitamin B_{12} is signalled by the appearance of hypersegmented polymorphonuclear neutrophils in the circulating blood with an increased lobe average.

FOLIC ACID

Folacin, the second to the last vitamin to be discovered, was found during the search for the factor in liver responsible for curing anaemia.

Lucy Wills, a British physician, studied the origins of a macrocytic anaemia prevalent among pregnant textile workers in Bombay. The anaemia was associated with poverty and a diet deficient in animal protein and vegetables. Yeast and liver extracts cured anaemic patients in Bombay (Wills, 1933). There was no neurological change in the human anaemia of pregnancy and poverty and deficient monkeys responded to a soluble fraction of liver or yeast that was ineffective in patients with pernicious anaemia.

BIOCHEMISTRY

The active form of folate consists of a substituted pteridine ring linked to *p*-aminobenzoic acid (PABA) (together forming pteroic acid) and glutamic acid. The term folic acid refers to pteroyl glutamate which is the monoglutamyl form of the vitamin. This vitamin occurs in food however with one to seven glutamic acid molecules attached.

Folic acid forms yellow crystals and is a conjugated substance made up of three acids.

FUNCTION

ᗢ *Basic coenzyme role* Folic acid coenzyme is a necessary agent in the important task of attaching single carbon to compounds such as CH_3 groups.

   a. Folic acid coenzyme is used in the synthesis of purines and is involved in cell division and in the transmission of inherited traits essential for DNA synthesis.

   b. Coenzyme is necessary for synthesis of thymine which is an essential component of DNA.

   c. It is necessary for the synthesis of haem which is an iron containing non-protein portion of haemoglobin.

▾ The principal interaction with vitamin B_{12} is in the action of methionine synthetase.

▾ Along with B_{12} the folic acid helps in transmethylation of homocysteine to methionine, ethanolamine to choline and uracil to thymine.

▾ Helps in the conversion of phenylalanine to tyrosine.

▾ The conversion of the vitamin niacin to *N*-methyl nicotinamide, the form in which it is excreted.

▾ Folic acid is required for the normal metabolic pathway of histidine, particularly in the conversion of formimino glutamic acid to glutamic acid.

Examples Folacin is important in the regeneration of RBC and of those cells lining the gastrointestinal tract where rapid cell division is occurring. The formation of each new cell requires the synthesis of DNA to carry its genetic information. Because nucleic acids control protein synthesis, folacin exerts an indirect effect on the synthesis of enzymes and other essential proteins.

The formation of the porphyrin group of haemoglobin before iron is added and the metabolism of long-chain fatty acids in the brain.

METABOLISM AND ABSORPTION

Digestion

Dietary folates exist as polyglutamated, reduced and substituted forms of pteroylglutamate (folic acid) which have to be hydrolysed to the monoglutamyl form. This usually occurs in the cells in the intestinal wall involving specific enzymes and vitamin B_{12}.

In humans, there are two folate hydrolases in the intestinal mucosa, one in the brush border membrane and the other in the intracellular lysosomes.

Brush border enzymes are responsible for the digestion of a majority of dietary folates.

Intestinal zinc from dietary sources is essential for polyglutamyl folate hydrolysis.

Absorption

Following hydrolysis, the monoglutamyl folate derivative is bound to a specific folate receptor on the brush border of the enterocyte. Folate is reduced and methylated intracellularly and reaches the liver via the portal circulation as 5-methyl H_4 folate. The total body folate pool size is estimated at 7.5 ± 2.5 mg for healthy adults.

The liver is the principal site of folate storage and contains about 7 pg folate/g.

Binding Proteins and Transport

Folate in plasma is distributed in three fractions, as free folate and that loosely bound to low-affinity binders, e.g. albumin and high-affinity binders (less amounts being bound), e.g. glycoproteins. Transport across cell membranes is as monoglutamates.

Storage is in the liver. Total-body folate stores range from 5 to 10 mg. Most stored folate is present as polyglutamates in the liver.

Excretion

Folate is excreted in urine and bile in metabolically active and inactive forms. About 100 mg of biologically active folate is excreted in the bile daily.

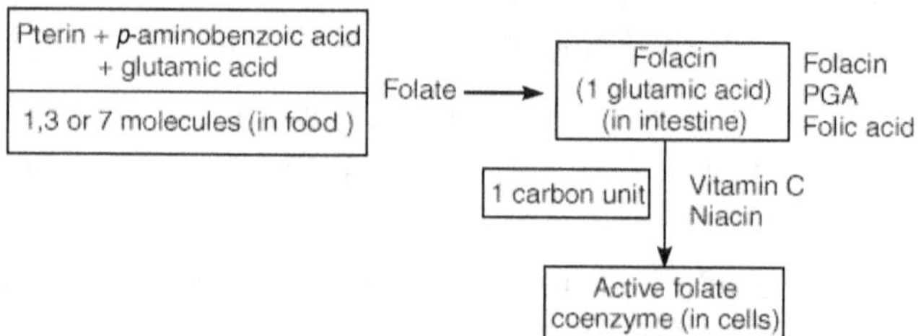

DEFICIENCY

Because folate is required for DNA synthesis, its deficiency is expressed in tissues with high rates of cell turnover. The principal sign of deficiency is megaloblastic anaemia. The aetiology of folate deficiency can be grouped into five categories.

1. *Decreased dietary intake*   Occurs among people ingesting inadequate or improper diet, alcoholics and the poor.

2. *Decreased intestinal absorption*   Intestinal malabsorption causes deficiency in patients with coeliac disease or tropical sprue.

3. *Increased requirements*   During pregnancy, lactation haemolytic anaemia and in leukemia.

4. *Effects of drugs*   Patients with inflammatory bowel disease receiving salicylazosulphapyridine are at risk of folate deficiency because this drug blocks both hydrolysis and intestinal uptake of dietary folate. Epileptic drugs also cause deficiency.

5. *Alcoholism*   Alcoholic liver disease increases likelihood of deficiency. Megaloblastic bone marrow changes and anaemia occur in a third of chronic alcoholics.

A deficiency of the vitamin has been implicated in pregnancy-induced hypertension. Folacin deficiency is associated with foetal damage and severe depletion of maternal reserves.

Folate deficiency is first seen in tissues which are actively dividing such as the intestinal cells and red blood cells. It has also been implicated in pregnancy-induced hypertension, proteinuria and oedema. Folate deficiency during pregnancy causes damage to the foetus, i.e., neural tube disorders. It may also induce the production of leucocytes and thus impair the ability of the immune system to fight infection.

There may be localized folate deficiency which causes megaloblastic changes in the cervix in women taking oral contraceptives. Such localized changes are also seen in bronchial and oral tissues of smokers and the colonic tissue of patients with ulcerative colitis.

Megaloblastic Anaemia

It is an orthochromic macrocytic anaemia, where the RBCs are irregular in size and shape usually larger than normal and have their full complement of haemoglobin (Figure 8.5). There is impaired DNA synthesis, cell maturation and division.

The symptoms are fatigue, diarrhoea, glossitis, parasthesia and associated with tropical sprue.

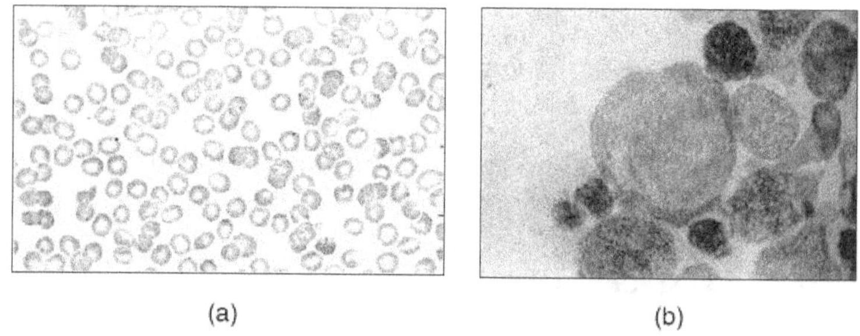

Figure 8.5   (a) Normal blood cells (b) Blood cells in macrocytic anaemia

Treatment   Folic acid doses of 5–10 µg daily and 1000 µg of vitamin B_{12} is given weekly as folic acid deficiency is associated with vitamin B_{12} deficiency.

FOOD SOURCES

Rich sources are wheat germ (178 µg/100 g) liver, kidney, yeast and mushrooms.

Good sources are fruit and vegetable, oranges and orange juice (17 to 43 µg/100 g), asparagus, broccoli and spinach, lemons, bananas, strawberries and cantalopes.

REQUIREMENTS

ICMR recommendations for dietary allowances of folic acid is given in the following table.

Man	100 µg
Women	100 µg
Pregnancy	400 µg
Lactation	150 µg
Infants	25 µg
Children	
1–3 yrs	30 µg
4–6 yrs	40 µg
7–9 yrs	60 µg
Adolescents	
10–12 yrs	70 µg
13–15 yrs	100 µg
16–18 yrs	100 µg

EVALUATION OF FOLATE STATUS

This can be done as follows:

- Identification of megaloblastic anaemia
- Plasma folate levels of < 3 ng/ml
- Elevated plasma homocysteine

VITAMIN C (ASCORBIC ACID)

Scurvy was reported as early as 1500 CE and in the writings of Hippocrates in 400 BC. It was known as early as the seventeenth century that scurvy could be controlled by eating certain food and since 1906 it was known that it is a vitamin deficiency disease.

Vitamin C was isolated in 1932 from lemon juice, oranges and cabbage.

CHEMICAL STRUCTURE AND PROPERTIES

Chemically, vitamin C is a simple compound of six carbon atoms, closely related to the monosaccharide, glucose.

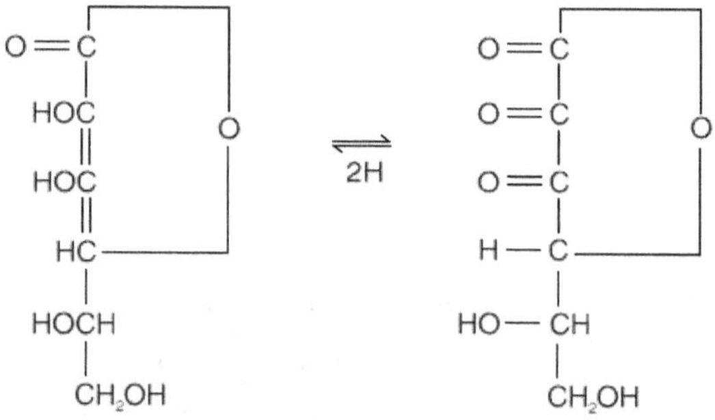

$$
\text{L-Ascorbic acid} \quad \rightleftharpoons^{2H} \quad \text{L-Dehydroascorbic acid}
$$

L-Ascorbic acid   L-Dehydroascorbic acid

It is a white crystalline substance which is stable when dry but easily oxidized in solution in water especially in an alkaline medium. It is stable to acid but easily destroyed by oxidation, light, alkali and heat, especially in the presence of iron or copper.

Ascorbic acid is a powerful reducing agent. It contributes electrons to other substances, thus having the same effect as an antioxidant.

Vitamin C, usually present in food as reduced ascorbic acid, is susceptible to oxidation, causing it to change on exposure to air to dehydroascorbic acid, which has two hydrogen atoms lesser. It is transported to and enters the cell more easily in this oxidized form and is apparently reduced again within the cell before it can be used. Any further oxidation of dehydroascorbic acid is irreversible, having produced a biologically inactive form, diketogulonic acid, with no vitamin value.

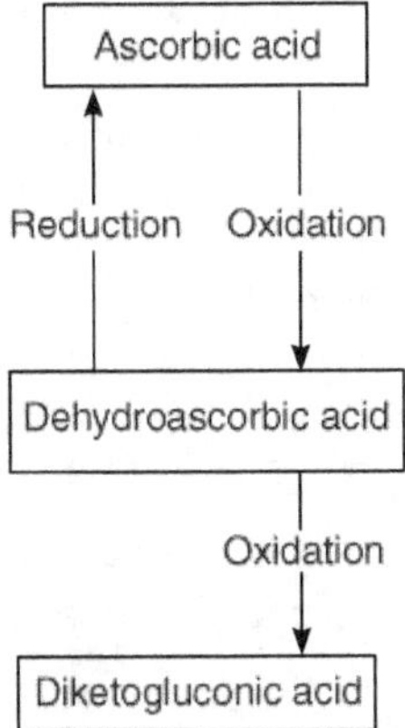

Vitamin C occurs in two forms-D-ascorbic acid and L-ascorbic acid, the latter predominantly in food. Although they differ only in the way the atoms are arranged, L-ascorbic acid is well used by humans; D-ascorbic acid, on the other hand, can be used only in small doses. Because it is used extensively as a preservative in processing meat, D-ascorbic acid is identified as erythrobic acid (used to preserve colour of meat but cannot function as a vitamin).

Synthesis   Most animals can synthesize vitamin C from glucose but humans lack the enzyme necessary to complete the conversion of glucose or galactose to vitamin C.

FUNCTIONS

1. **Collagen formation**   Collagen is a major component of all connective tissue, where it binds cell together. It is also a major component of skin, cartilage, teeth and scar tissue, and it provides structural framework of bone.

   Vitamin C functions in collagen formation by promoting the change of the amino acids lysine and proline in strands of tropocollagen to hydroxylysine and hydroxyproline which are necessary parts of collagen fibres. This process is called hydroxylation. This must take place to make mature collagen fibre, which in turn makes strong connective tissue.

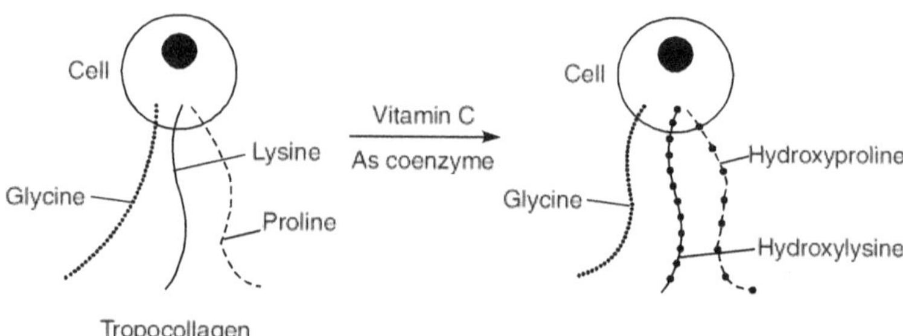

This same process is involved in the formation of scar tissue. Skin grafts for burned tissue heal more quickly when vitamin C is present.

Failure of collagen synthesis also causes small pinpoint haemorrhages resulting from weakness in the membranes that line the capillaries and in the fibres that join the cells together under the surface of the skin. These subcutaneous haemorrhages show up most often in areas subjected to mechanical stress such as the gums.

The bone matrix which is primarily collagen, is less capable of holding calcium and phosphorus during bone calcification resulting in weakened or distorted bone structure during vitamin C deficiency.

Ascorbic acid is a reductive cofactor for post-translational hydroxylation of peptide-bound proline and lysine residues during formation of collagen. The hydroxyproline and hydroxylysine units allow cross-linking to stabilize the triple helical structure of tropocollagen. The reaction occurs in the endoplasmic reticulum of the fibroblast before excretion of procollagen from the cell. The enzyme involved in proline hydroxylation, prolyl hydroxylase, requires molecular oxygen, ascorbic acid, iron and α-ketoglutarate.

Ascorbic acid may also serve as a reductant for diverse metal-dependent polymerization and cross-linking reactions of connective tissue, and as a carrier for sulphate groups needed for production of glycosaminoglycans.

Peptide-bound proline + α-ketoglutarate →[Prolyl hydroxylase, Fe and vitamin C]→ Peptide-bound hydroxyproline + Succinate + CO_2

2. **Carnitine biosynthesis**   Carnitine is required as a transporter of long chain fatty acids across the mitochondrial membrane wherein β-oxidation provides energy to cells, especially for cardiac and skeletal muscles. The biosynthesis involves the methylation of lysine with methionine as methyl donor and requires ascorbic acid, iron, vitamin B_6 and niacin as cofactors of various enzymes of the pathway.

3. **Dentin formation**   The dentin layer of the tooth, derived from a group of cells known as odontoblasts, does not form normally when vitamin C is lacking. This produces a tooth with structural weakness that is less able to resist mechanical injury or decay.

4. **Tyrosine metabolism**   Vitamin C is necessary for hydroxylation of large amounts of the amino acid tyrosine. Because tyrosine is a precursor of the hormone thyroxine and the neurotransmitter norepinephrine, vitamin C is indirectly involved in thyroid and adrenal function.

Tyrosine → Dopamine →[Dopamine-β-hydroxylase, Cu and ascorbic acid]→ Noradrenaline

Ascorbic acid appears also to be involved in the hydroxylation of tryptophan to form serotonin in the brain and in the degradation of tyrosine by *p*-hydroxyphenyl pyruvate hydroxylase.

5. **Synthesis of neurotransmitters**   In the brain two of the neurotransmitters needed to transfer nerve impulses from one cell to another can be produced only if adequate levels of vitamin C are available. Vitamin C is needed to convert tyrosine to the neurotransmitter norepinephrine and the amino acid tryptophan to the precursor of the neurotransmitter serotonin.

   Ascorbic acid is required as a cofactor for the copper-containing dopamine-β-mono-oxygenase enzyme that catalyses hydroxylation of the dopamine side chain to form norepinephrine.

6. **Utilization of iron, calcium and folacin**   Vitamin C readily gives up an electron to convert Fe^{3+} to Fe^{2+}. This conversion takes place in the lumen of the gut and helps in absorption of Fe (as it can be absorbed only in the ferrous form). It also helps in the transfer of iron from the blood into ferritin to be stored in the liver and activates some iron-containing enzymes. Similarly vitamin C aids calcium absorption by keeping it from forming an insoluble complex.

   The conversion of the inactive form of the vitamin folacin to the active form is facilitated by vitamin C. It also increases the stability and use of folic acid and vitamin E.

Other Functions

1. Ascorbic acid is also necessary for other functions which include the following.

   i. for the degradation of cholesterol to bile acids.

   ii. for detoxification of drugs in the liver.

   iii. it is suggested that vitamin C has a role in cancer control. As a reducing agent it could prevent the oxidation of harmless precursors to carcinogens such as the conversion of nitrates to nitrites, the precursor of nitrosamines; it could promote the synthesis of mucopolysaccharides (ground substance) which inhibit the growth of cancerous cells; or it could provide protection against the stress of surgery, chemotherapy, or radiotherapy.

iv. Vitamin C helps to detoxify histamine thus alleviating symptoms of many conditions like hay fever, frostbite or poisoning.

2. **Mixed function oxygenase system**   The microsomal drug-metabolizing system operates in liver microsomes and reticuloendothelial tissues to inactivate and metabolize a wide variety of substrates such as endogenous hormones or xenobiotics. The systems operate with oxygenase enzymes, flavoproteins, cytochrome P450 protein, oxygen and reducing agents such as NAD(P)H. The activity of these systems depends on ascorbic acid.

Vitamin C is also involved in the hepatic microsonal hydroxylation of cholesterol in the excretion of cholesterol as bile acids.

3. **Cyclic nucleotides**   Ascorbic acid increases cellular levels of cyclic adenosine monophosphate and cyclic guanosine monophosphate (cAMP and cGMP) by increasing their synthesis.

4. **Miscellaneous functions**   Vitamin C, as ascorbate 2-sulphate, serves as a sulphating agent of cholesterol as part of its catabolism and of mucopolysaccharides during formation of connective tissue.

Ascorbic acid also stimulates prostaglandin synthesis.

ABSORPTION AND METABOLISM

Vitamin C is absorbed in the upper part of the intestines either by simple diffusion or by a sodium-dependent active transport mechanism (requires energy and depends on an exchange of sodium from within the cell).

Vitamin C is stored in organs like adrenal gland, kidney, lung, liver, blood and muscles. The total pool of vitamin C in the body reaches 1500 mg. Vitamin C in the serum reaches a maximum of 1.2 mg/L. Vitamin C is excreted primarily in the urine.

DEFICIENCY

Scurvy, the most severe form of vitamin C deficiency is relatively rare (Figure 8.6). The early symptoms are listlessness, fatigue, weakness, shortness of breath, muscle cramp, aching bones, joints and muscles and loss of appetite. The skin becomes dry, feverish and rough and is covered by reddish blue spots. Haemorrhaging of gums is seen. Enlargement or hypertrophy of cornea, congestion of follicles, difficulty in breathing are also seen. Petechiae, or pinpoint haemorrhages under the skin develop.

Infantile scurvy develops in infants fed only with cow's milk and whose diets are not varied by six months and symptoms such as irritability, anorexia, growth failure tenderness of lips and anaemia are seen. The onset is extremely rapid and unless treated promptly may result in death.

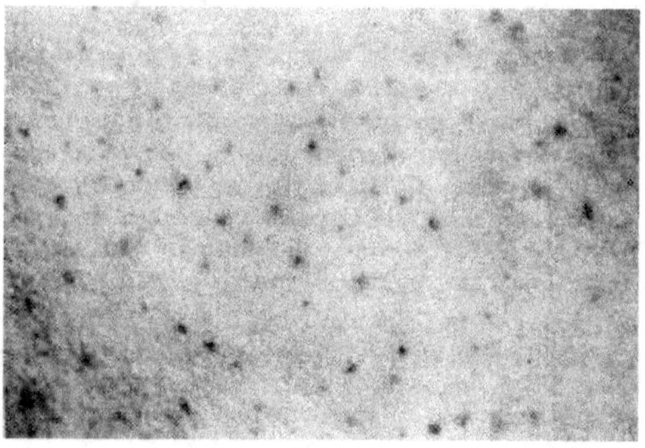

Figure 8.6   Pinpoint haemorrhages on the skin—an early stage of scurvy

Delayed or incomplete wound healing is a frequent sign of ascorbic acid deficiency and invariably anaemia occurs after 2 months of restricted intake.

Vitamin C and smoking   Smokers have an increased need for vitamin C. They absorb 10% less.

DEFICIENCY

Manifestation

Mesenchymal manifestation is as follows:

- Petechiae
- Ecchymose
- Coiled hairs
- Perfollicular haemorrhages
- Inflamed and bleeding gums
- Hyperkeratosis
- Dyspnea
- Joint effusions
- Arthralgia
- Oedema
- Impaired wound healing

Systemic manifestation is as follows:

- Weakness
- Fatigue
- Lassitude

Psychological and **neurologic** manifestation is as follows:

- Depression
- Hysteria
- Hypochondriasis
- Vasomotor instability

FOOD SOURCES

Vitamin C is found exclusively in foods of plant origin. Liver contains a small amount.

To help retain vitamin C in food during preparation the following are maintained.

1. Harvest at peak of maturity.

2. Store in a cool moist place.

3. Limit exposure to air.

4. Do not soak in water.

5. Cook in a small amount of water.

6. Leave food in large pieces if possible.

7. Use microwave oven for cooking.

   microwave 15% loss

   steaming 30%

   pressure cooking 20%

   boiling 55%

Gauva, lime, oranges, cabbage, broccoli, strawberries, tomato, potato, etc. are fair sources of vitamin C.

REQUIREMENTS

Children	
1–10 years	45 mg
11–14 years	50 mg
Men	
15+ years	60 mg
Women	
15+ years	60 mg
During pregnancy	+20 mg
During lactation	+40 mg

CLINICAL, PROPHYLACTIC AND THERAPEUTIC ASPECTS OF VITAMIN C

1. **Immunocompetence** A variety of immune-related functions have been associated with ascorbic acid status including decreased resistance to a variety of infections agents during ascorbic acid deficiency and effects on neutrophil activity and lympocyte blastogenesis. Stimulation of neutrophil chemotaxis is suggested as the mechanism underlying improved antimicrobial activity.

2. **Wound healing** Ascorbic acid deficiency delays wound healing as collagen formation is affected. The rapid wound healing requires the formation of strong connective tissue in the scar.

3. **Antioxidant defence** Ascorbic acid scavenges aqueous superoxide and hydroxyl radicals and acts as a chain-breaking antioxidant in lipid peroxidations. It also acts indirectly in protecting lipid membranes by regenerating the active form of membrane-bound vitamin E.

   High levels of ascorbic acid in the humor of the eye may provide antioxidant protection to various fluids and tissues, including lens, cornea, vitreous humor ocular and retina.

4. **Cancer** Anticarcinogenic effects of ascorbic acid is due to its ability to detoxify carcinogens or block carcinogenic processes (through antioxidant or free-radical scavenging actions) and to enhance immunocompetence. Studies have shown that people who eat more fruits and vegetables do not develop certain types of cancer as they have a higher intake of vitamin C.

5. **Lipid metabolism and heart disease** Ascorbic acid prevents heart disease; vascular tissue integrity and reduces thrombotic episodes. It plays a protective role in the prevention of atherosclerosis.

6. **Other clinical aspects** Increased ascorbic acid levels helps lower body histamine levels and hence helps patients with asthma and hypersensitivity.

Ascorbic acid has been shown to ameliorate heavy metal toxicity by reducing metal absorption or converting metals to less toxic forms.

7. **Common cold and flu**   The cause of the common cold and flu is modified by vitamin C intake.

REVIEW QUESTIONS

1. Give the functions of thiamine.
2. What is beriberi?
3. Explain the role of B vitamins in metabolism.
4. Give the sources of thiamine.
5. Give the RDA of thiamine.
6. What is pellagra?
7. How can losses of B vitamin be prevented during cooking?
8. What is glossitis, cheilosis and angular stomatitis.
9. Give the functions of folic acid.
10. What is pernicious anaemia?
11. Give sources of folic acid and vitamin B_{12} for different age groups.
12. Why is the body's need for vitamin B_6 related to protein intake?
13. Discuss deficiency symptoms of B_6.
14. List the functions of pantothenic acid, biotin and B_6.
15. Give the RDA for biotin.
16. List the functions of vitamin C.
17. Give the symptoms and cure for scurvy.
18. Give the best sources of vitamin C.
19. Write a note on the antioxidant property of vitamin C.

CRITICAL THINKING QUESTION

1. Can vitamin C prevent common cold?

MACRO MINERALS

Did You Know?

- Calcium makes up the largest amount of the total body weight.

- Calcium is necessary for bone and teeth formation.

- It is also important for muscle contraction and relaxation.

- Phosphorus is found in combination with calcium.

- A Ca : P ratio of 1 : 1 promotes absorption.

- Milk is a good source of calcium.

CALCIUM

DISTRIBUTION

Of all minerals, calcium makes up the largest, i.e., between 15% and 20% (850 to 1200 g) of the total body weight, most of it (99%) being present in bones and teeth. The rest is widely distributed in the blood and soft tissues such as muscle, liver and heart. In blood, Ca levels are maintained within a very narrow range of 9 to 11 mg/dl with about half loosely bound to protein and almost all the rest present as free calcium ions. A very small amount (4–5 g) is vital to the functioning of every body cell. Calcium in the blood and soft tissues is so important that it will be mobilized from reserves in the bones to maintain constant blood levels.

The bulk of skeletal calcium is deposited as hydroxyapatite, $Ca_{10}H(PO_4)_6(OH_2)$. This bone tissue is constantly being reshaped according to various body needs and stresses, with as much as 700 mg calcium entering and leaving the bones each day.

The level of calcium in blood is strictly guarded by 2 main hormones;

1. Vitamin D hormone (1, 25 dihydroxycholecalciferol)

2. Parathyroid hormone

This small but highly significant amount of calcium in plasma and other body fluids occurs mainly in 2 parts, non-diffusible and diffusible, with a small bit as part of organic complexes. These 3 parts are normally in equilibrium.

1. **Non-diffusible calcium** About half of the plasma calcium is bound with the plasma proteins, albumin and globulin.

2. **Diffusible ionized free calcium (Ca^{++})** This makes up the other half of the calcium in plasma. It has the greatest physiologic effect among the 3 fractions, exerting a

profound influence on metabolism and functions of bones, nervous system and the heart.

3. **Organic complexes**   About 7% of the plasma Calcium is diffusible but occurs as a part of organic complexes such as citrate and other substances.

The distribution of calcium in blood is shown in Figure 9.1.

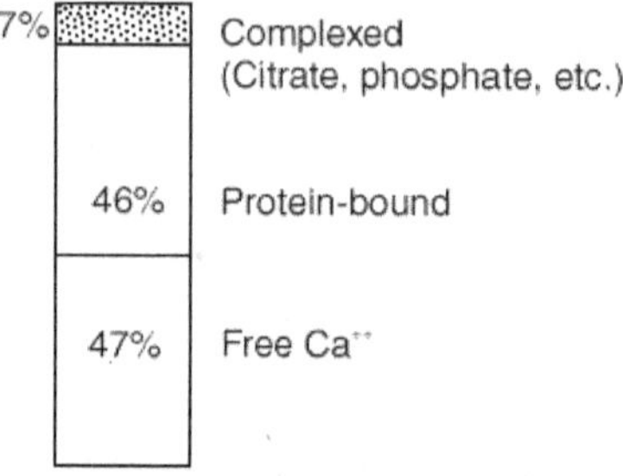

Figure 9.1   Distribution of calcium in blood

Calcium in human body

Skeleton	
Young adult male	1200 g
Newborn	30 g
Blood and extracellular fluid	8.5–10.5 mg/dl
Cells and intracellular fluid	8 mg/dl
Daily turnover in adult skeleton	400–600 mg
Daily urinary output	50–350 mg
Daily faecal output	650–900 mg

FUNCTIONS

Bone Formation

Early in foetal development, a strong but flexible matrix, or pattern, for bone made of the protein collagen is formed. It has the same general shape as the mature bone but lacks strength and rigidity. The matrix, which accounts for one-third of the bone, remains rather flexible until after birth (possibly making the birth process easier). It is composed of fibres of the protein collagen embedded in a gelatinous substance. After birth, the matrix starts to gain strength and rigidity primarily as a result of growth of mineral crystals in the matrix in a process known as calcification or ossification, i.e, hydroxyapatite crystals are deposited on the matrix. The bone shaft is constantly remodelled and reshaped throughout life as body weight shifts (Figure 9.2).

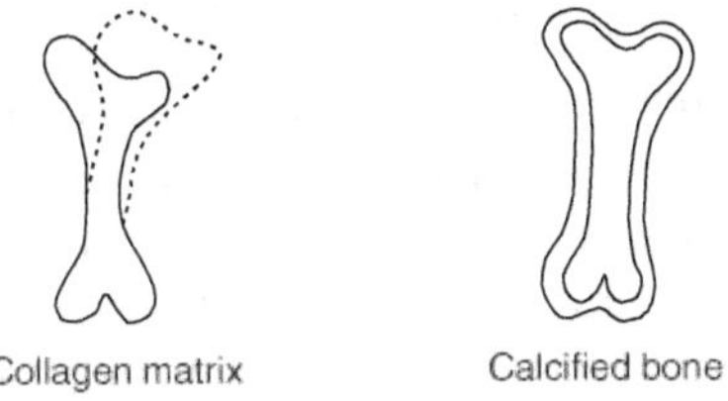

Figure 9.2   Bone formation

The process of building and maintaining skeletal tissue is a delicately balanced process which is carried on by 2 types of cells. Osteoblasts continually form new bone matrix in which calcium phosphate is deposited and bone crystals develop.

Osteoclasts continually balance this activity by absorbing bone tissue. Osteoclasts phagocytoze, i.e., digest minute bone crystals.

Tooth Formation

Hydroxyapatite, the mineral compound found in bones, is the same one that forms dentin and enamel, which are the middle and outer layers of the tooth. However the crystals in teeth are more dense and the water content is lower. The protein in enamel is keratin whereas that in dentin is collagen. Tooth-forming organs in the gum, the ameloblasts, deposit calcium and other constituents to form the teeth. Permanent teeth have no ability to repair themselves once erupted. A deficiency of calcium during the formative period may lead to increased susceptibility to tooth decay.

Growth

A Japanese study showed that people raised on a diet low in calcium are frequently shorter than those given adequate calcium. This could be due to the fact that calcium is necessary for normal growth especially for bones.

General Metabolic Functions

1% of body's calcium is involved in metabolic functions.

Blood clotting Once cells have been injured, ionized calcium in the blood stimulates the release of a phospholipid, thromboplastin, from the injured blood platelets which catalyses the conversion of prothrombin to thrombin, which in turn helps activate fibrinogen to fibrin. Calcium ions are also required for cross-linking of fibrin, giving strength to fibrin threads.

Nerve transmission Calcium is required for normal transmission of nerve impulses. A current of calcium ions triggers the flow of signals from one nerve cell to another and on to the waiting target muscles. The calcium ions in the extracellular fluid at the neuromuscular junction cause the neurotransmitter acetylcholine to rupture through the separating membranes at the tips of many nerve branches and execute the muscle fibre.

Muscle contraction and relaxation Each muscle fibre is made up of hundreds of contractile units called myofibrils which are composed of the protein filaments myosin and actin. Calcium is bound to the tubular reticulum of the muscle. On activation, calcium is released, ionized and mobilized. The free calcium ions activate the chemical reaction between myosin and actin filaments and release large amounts of energy from ATP, thus bringing about contraction. When they are bound back to the reticulum they bring about relaxation. Calcium is important for the contraction and relaxation of the heart muscle. The tissue fluid that bathes heart muscle must contain adequate amounts of calcium in proportion to Na, K and Mg. If calcium salts are removed, muscles fail to contract. If calcium levels are high, a state of continuous contraction occurs and calcium rigour can occur (spasmodic contractions).

Cell membrane permeability Ionized calcium controls the passage of fluid through cell membranes by affecting cell wall permeability. It influences the integrity of the intercellular cement substances.

Enzyme activation Calcium functions as a catalyst or cofactor in many biological reactions. Among these reactions are:

   i.   absorption of B_{12}

   ii.   action of the fat-splitting enzyme, pancreatic lipase

   iii.   secretion of insulin by pancreas

   iv.   formation and breakdown of acetylcholine

   v.   calcium regulates contractility of muscles and delays fatigue of muscle by increasing the muscle O_2 content. In rats it was found that deficiency of calcium produces muscle weakness, paralysis of hind limbs, abnormal posture, decreased lifespan and is high mortality rate.

Increased lifespan It is been observed that people who consume larger quantity of milk and hence large amounts of calcium have a greater lifespan.

ABSORPTION

About 10–30% of calcium in an average diet is absorbed. Most food calcium occurs in complexes with other dietary constituents, which must be broken down and the calcium released in soluble form before it can be absorbed. Most absorption takes place in the small intestine chiefly in the duodenum where pH is lower than the distal portion of intestine. Calcium salts are insoluble in a less acid medium. A calcium-binding protein (CaBP) carrier is synthesized in the duodenum in response to the action of vitamin D hormone. Thus transport of calcium is by active transport, an energy-requiring process, across the cell through the membrane and is released into the blood.

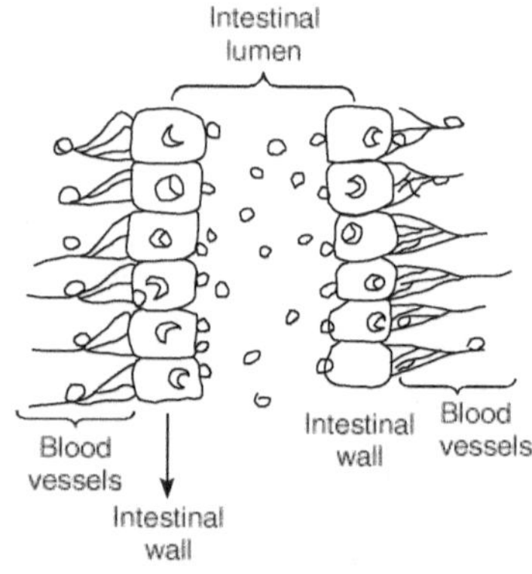

Figure 9.3   Active transport of calcium

In active transport (Figure 9.3), a calcium carrier shuttles calcium across the intestinal wall from the lumen to the blood.

Factors Favouring Calcium Absorption

1. **Vitamin D**   Vitamin D is changed to an active form which catalyses the synthesis of a protein, which serves as a carrier for calcium in the cells lining the intestinal wall. This active form can result in a 10–30% increase in calcium absorption.

2. **Need for calcium**   Absorption is influenced by the body's need for calcium. During pregnancy, lactation and adolescence when the needs for calcium are greatest, absorption rates are as high as 50%.

3. **Acidity of digestive mass**   Calcium is more soluble in acid and is therefore more readily absorbed from an acid medium than from an alkaline medium. Secretion of HCl is decreased in old age thereby decreasing calcium absorption.

4. **Presence of sugars**   Lactose enhances the absorption of calcium in the ileum. Other sugars such as ribose and fructose also enhance calcium absorption but lactose is the most effective. Recently it has been shown that 10 g of glucose also enhances calcium absorption as much as 25%.

5. **Dietary protein**   The favourable effect of protein in enhancing calcium absorption and reducing calcium losses in the urine has been attributed to both the phosphorus in protein and specifically to the amino acids arginine and lysine.

6. **Plasma ion concentration**   Even small decreases in the concentration of ionized calcium in the body fluids are reflected in a rise in calcium absorption. The calcium ion concentration in turn is mediated by an intracellular receptor protein, calmodulin, that binds calcium ions when their concentration increases in response to a stimulus.

7. **Calcium/phosphorus ratio**   A ratio of Ca : P of 1 : 1 to 1 : 2 (avg. 1 : 1.5) promotes absorption. Diets should have a Ca : P ratio of 1 : 1 during rapid growth and calcification, when calcium is required only for maintenance as in adults, the calcium requirement is lower than the phosphorous requirement. The Ca : P ratio is highest at the earliest age (1 : 1) and decreases with attainment of adult status 1 : 2 and then in the case of females increases again (1 : 1) in the latter half of pregnancy and lactation. In normal adults the Ca : P ratio of the whole body is 1 : 1 and bone is 2 : 1.

If either mineral is taken in excess of this ratio, absorption of both is hindered and excretion of the lesser mineral is increased, e.g. excess phosphorous in relation to calcium results in formation of more $CaPO_4$ which binds calcium and makes it unavailable for absorption.

The amounts of calcium and phosphorus in the serum are maintained in a definite relationship called serum Ca : P ratio (this ratio is the product of Ca × P expressed in mg/dl).

Normal serum levels of Ca and P are

Ca   10 mg/dl

P   4 mg/dl (adults)

      5 mg/dl (children)

Factors Depressing Calcium Absorption

1. **Oxalic acid**   It is found in some green leafy vegetables such as spinach. It combines with calcium to form an insoluble compound, calcium oxalate. The calcium in calcium oxalate cannot be freed to be absorbed.

2. **Phytic acid**   This is another organic acid which binds calcium and makes it unavailable for absorption. It is found on the outer husk of cereals.

3. **Presence of dietary fat**   High fat diets promote the formation of insoluble soaps of fatty acids and calcium; they can result in steatorrhoea (fatty stools) and a reduction in calcium absorption.

4. **Emotional instability**   Emotional states such as stress, tension, anxiety, grief or boredom can interfere with calcium absorption.

5. **Increased gastrointestinal mobility**   Anything that increases the rate of passage of food through the intestinal tract decreases calcium absorption by reducing the amount of time that the contents of the intestinal tract are in contact with the intestinal wall.

6. **Lack of exercise**   People who do not exercise or are bedridden lose as much as 0.5% of bone calcium in a month and have a reduced ability to replace it.

7. **Dietary fibre**   The absorption of calcium is decreased either by decreasing the time needed to pass food through the

digestive tract, by increasing the bulk of the food mass, by diluting intestinal contents or by binding the mineral.

8. **Caffeine** High intake of caffeine affects calcium bioavailability by increasing the loss of calcium in the urine and the secretion and loss of calcium through the gastrointestinal tract.

9. **Drugs** Some medications such as anticonvulsants, cortisone and thyroxine reduce calcium absorption.

EXCRETION

Dietary calcium in excess of need remains unabsorbed in the intestine and is excreted in the faeces. Calcium is also excreted by way of kidney and skin (Figure 9.4).

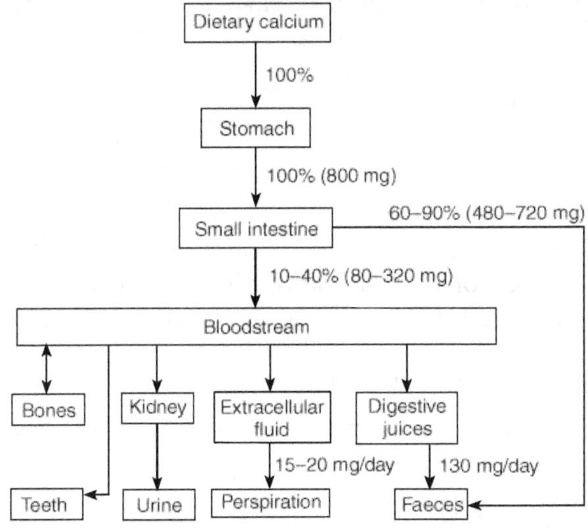

Figure 9.4   Excretion of calcium

The various forms of calcium excreted through intestine are

1.   unabsorbed calcium

2. calcium present in epithelial cells

3. calcium present in digestive juices

On a 1 g intake in man, excretion is 622 mg (mean) and the daily urinary excretion is 100–175 mg. Loss through perspiration is 15–20 mg, but if there is profuse sweating, about 149 mg is lost.

Urinary calcium is on an average not influenced by daily intake but is related to the process of reabsorption in the kidney tubules. Excretion differs in individuals but there is constancy within each person.

METABOLISM

Calcium Homeostasis

The hormonal mechanism of calcium homoeostasis operates at 3 sites.

1. Bones

2. Kidney

3. Gut

PTH, calcitonin and metabolites of vitamin D help in the maintenance of calcium homeostasis. The following helps to understand maintenance of serum calcium levels.

I. 1. Fall in serum calcium

2. Parathyroid gland is stimulated—parathormone

3. Parathormone stimulates biosynthesis of 1, 25 DHCC

4. 1, 25 DHCC stimulates intestinal calcium absorption.

5. 1, 25 DHCC + DTH stimulates renal absorption of calcium and mobilization of calcium from bone and increased phosphorus excretion by kidneys and resorption by bone.

6. So rise in serum calcium

II. 1. Rise in serum calcium above normal

2. Thyroid glands secrete the hormone, calcitonin

3. Blocks mobilization of calcium from bone

4. Also stimulates reabsorption of calcium in bone

5. Stimulates calcium excretion

The above is summarized in Figure 9.5.

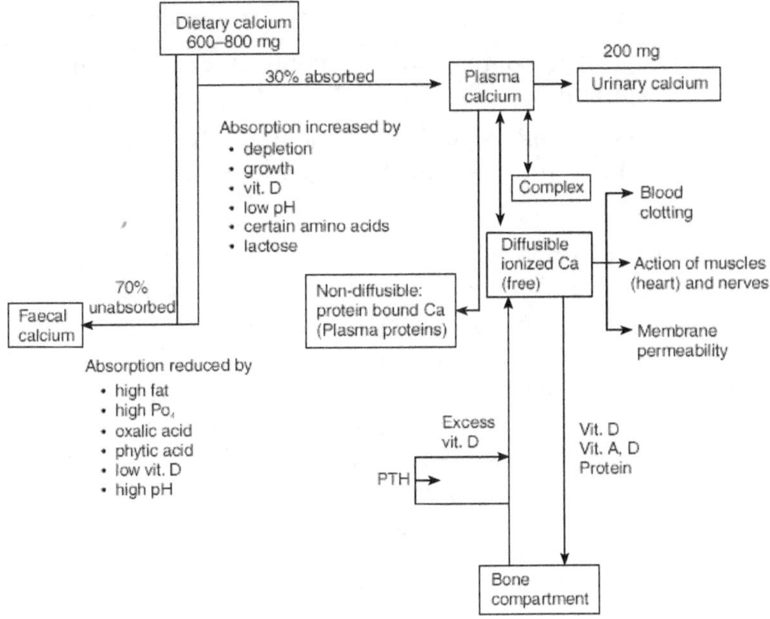

Figure 9.5   Calcium metabolism and functions

PTH does not directly influence calcium absorption but stimulates calcium absorption by stimulating 1, 25 DHCC. Both PTH and 1, 25 DHCC must be present to stimulate the mobilization and retention of calcium. The action of PTH is in minutes but 1, 25 DHCC's action requires many hours. Thus short-term control of serum calcium is by PTH and existing

1, 25 DHCC, while longer stimulation cause 1, 25 DHCC to increase thus stimulating intestinal absorption of calcium.

Normal serum ratio for adults is 40.

Calcium Metabolism

Four mechanisms are involved in maintaining the plasma calcium level within its narrow normal range.

1. Intestinal absorption–excretion balance
2. Renal adjustment of excretion (the kidney threshold)
3. Storage compartment—maintenance of calcium stores in bone
4. Hormonal regulation—calcium balance controlled by three interrelated hormones, vitamin D, PTH and calcitonin.

Glucocorticoid excess leads to bone loss, especially of trabecular bone, in Cushing's disease or when steroids are used to treat various diseases. The symptoms are delay in growth and skeletal maturation in children and osteoporosis in adults. The main effect is an inhibition of osteoblastic activity. There is a reduced incorporation of sulphate into cartilage and of amino acids into collagen. Glucocorticoids also impair both the active and passive transport of calcium through intestinal cell. Thyroid hormones stimulate bone resorption and both compact and trabecular bone are lost in hyperthyroidism. Hypothyroidism impairs bone-mobilizing effect of PTH. The circulating levels of $1,25(OH)_2 D_3$ tends to be higher, and both the intestinal absorption and renal reabsorption of calcium are increased. Growth hormone stimulates the growth of cartilage and bone. It also stimulates 25(OH) hydroxylase increasing $1,25(OH)_2 D$ levels and the active intestinal transport of calcium. Insulin stimulates the osteoblastic production of collagen and directly reduces the renal reabsorption of calcium and sodium.

Oestrogen is essential for the maintenance of calcium balance in the bone. Bone contains oestrogen receptors which prevent bone resorption.

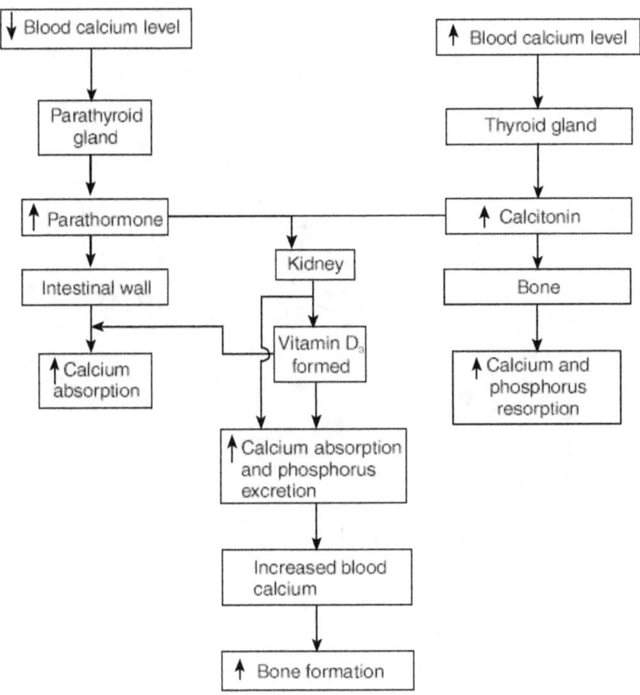

Figure 9.6   Hormonal regulation of calcium

Hormonal Regulation

Although PTH, calcitonin and vitamin D are the major calcium-regulating hormones, several other hormones affect bone turnover and calcium metabolism. These include glucocorticoid thyroid hormone, growth hormone, insulin and oestrogen (Figure 9.6).

RECOMMENDED DIETARY ALLOWANCE (RDA)

It is evident that special conditions create increased calcium needs, such as low birth weight infants, pregnant and lactating women, the elderly individuals with diseases involving intestinal malabsorption and patients with intestinal surgery.

Infants (Birth–1yr)	500–600 mg
Children (1–12 yrs)	400–700 mg
Adolescents (13–18 yrs)	600–700 mg
Men	400–500 mg
Women	400–500 mg
Pregnancy (latter half)	1000 mg
Lactation	1200 mg

DIETARY SOURCES

Rich sources   Milk and milk products, sesame seeds with husk, small dried fish

Fair sources   Egg, green leafy vegetables, ragi, broccoli, nuts, legumes

DISEASES OF CALCIUM DEFICIENCY OR EXCESS

Hypocalcaemia

The following are the factors that lead to hypocalcaemia.

1. Hypoparathyroidism causing diminished mobilization of Ca^{++} from bones.

2. Impaired elementary absorption due to vitamin D deficiency.

3. Malabsorption.

Tetany

A decrease in ionized serum Ca^{++} below a critical level results in changes in stimulation of nerve cells resulting in increased excitability of nerve and uncontrolled contraction of muscle called Tetany.

Tetany is characterized by severe, intermittent spastic contractions of the extremities and muscular pain. There is characteristic carpopedal spasm of the muscle in the upper extremity, which causes flexion of the wrist and thumb with extension of the fingers—Trusseau's sign.

Tetany-like responses may be caused by an increase in serum fraction of the calcium–phosphorus ratio, which causes a decrease in Ca^{++} level to maintain the ratio, e.g. milk. Tetany occurs in newborn infants fed undiluted cow's milk.

The ratio of phosphorus to calcium in cow's milk is greater than human milk, and the kidneys of the infants cannot clear this phosphate load which accumulates in serum. The rise in serum phosphorus causes a compensatory decrease in serum Ca^{++} and this in turn causes tetanic muscular spasms. When calcium level in blood rises above normal, the muscles are in a constant state of contraction called calcium rigor. Neither of these conditions is the result of abnormal dietary levels of calcium but due to improper functioning of parathyroid gland.

Osteoporosis

The rate of bone formation is normal in osteoporosis but bone resorption occurs at an accelerated pace leading to a decrease in bone mass. It is caused by

1. Decreased dietary Ca^{++}

2. Poor absorption or utilization of calcium

3. Irregularities of parathyroid gland in stimulating bone resorption

4. Immobility

5. Loss of stimulus for bone formation provided by oestrogen

Heredity, age, smoking, level of physical activity and certain medications are other causes of osteoporosis.

In most patients, osteoporosis is not corrected by decreased calcium alone but is often improved by exercise coupled with calcium supplements. In osteoporosis, the mass amount of bone in the skeleton has been diminished but the remaining bone mass is of normal composition. Bone fracture is common among osteoporotics. People who develop osteoporosis have a lower than normal intake of calcium over a long period of time and is more common in middle aged and elderly women. A diet high in calcium (15.5 mg/kg body wt.) with 50 mg/day of sodium fluoride and with adequate vitamin D has been shown to arrest the resorption of bone. It has been observed that the best way to prevent osteoporosis is with an adequate calcium intake throughout life. Fluoride may protect against bone mass loss by stimulating bone formation and by producing a more stable mineral.

More adequate intake of calcium throughout adult life, either from food, supplements or both help to maintain a larger bone mass loss to reduce the possibility of osteoporosis. Once the bone mass has diminished, however, supplements of calcium have no apparent effect in restoring the lost bone, they merely depress further loss.

Osteomalacia

It is the quality and not the quantity of bone that is reduced in osteomalacia. Although osteomalacia is associated with low phosphorus levels, low blood calcium levels is the most frequent cause of this disease. It is more common in

- Women living in areas of low sunshine

- Those taking anticonvulsant drugs

- Women with depleted mineral reserves resulting from demands of successive pregnancies and prolonged lactation. Most people with osteomalacia respond to vitamin D therapy.

Hypercalcaemia

Hypercalcaemia occurs under the following conditions.

- When there is too much of calcium in the blood.

- When baby foods supplemented with vitamin D are given.

- Hypercalcaemia is corrected by reducing vitamin D intake rather than lowering the amount of calcium in the diet.

Blood Pressure

Low calcium intake (<400 mg/day) increases blood pressure.

When calcium intake increases, there is a decrease in blood pressure.

PHOSPHORUS

Phosphorus is an important mineral constituent. Phosphorus is found in the bones as hydroxyapatite $[Ca_{10}(PO_4)_6(OH)_2]$.

Phosphorus in the extracellular fluid accounts for only 1% of the total phosphorus in the body. The majority (70%) of the total phosphorus in human plasma is found as a constituent of organic phospholipids. 10% of the circulating inorganic phosphorus is bound to protein, 5% is complexed with calcium or magnesium and the majority of plasma inorganic phosphorus is found as $H_2PO_4^-$ and HPO_4^{2-}. These two orthophosphate fractions are found at pH 7.4 in the ratio 1 : 4; plasma phosphorus is about 1.8 mEq/L.

METABOLISM

Digestion

Insoluble mineral PO_4 salts are formed at high alkaline pH. The stomach (pH = 2) and proximal part of the small intestine (pH = 5) play an important role in maintaining the luminal solubility and bioavailability of phosphorus. In vegetarian diets, a large part of the dietary phosphorus may be in the form of phytate. Hence, it may be unavailable for use. However, the traditional use of yeast in breadmaking liberates some phosphorus from the phytate moiety. The hydrolytic action of yeast phytases prior to baking may release some phosphorus, and intestinal bacteria normally located in the large intestine can breakdown some dietary phytate and hence may be available in small amounts. Other organic forms of phosphorus in the diet are primarily derived from typical cellular phosphorus-containing compounds such as phospholipids and phosphorylated sugars. Such compounds are digested in the intestine to liberate inorganic phosphate which is transported across the cell.

Absorption

About 60–70% of phosphorus is absorbed from a typically mixed diet. Phosphorus absorption is linearly related to phosphorus intake. Absorption is with the help of vitamin D. Physiological conditions like growth, pregnancy and lactation increase phosphorus need and hence increase phosphorus absorption.

Transport of phosphorus across the intestinal cell wall is by an active, sodium-dependent pathway, i.e., a carrier-mediated mechanism. Intracellular phosphorus level is high and the cell interior is electronegative, hence active transport is necessary to get the phosphorus into the cell, but the phosphorus may exit the cell by diffusion. A sodium sensitive-polypeptide has been identified at the brush border as a phosphorus transporter 1, 25 DHCC increases phosphorus absorption.

Excretion

The kidney is the primary route of phosphorus excretion. The transport of phosphorus in the renal tubules occurs by two processes. In the proximal convoluted tubule (PCT) of the kidneys, an active Na-dependent phosphorus transport is found.

STORAGE

Phosphorus is found in all cells and is the sixth most abundant element in the body. The total body content of phosphorus is about 16 mol (500 g) in males and 13 mol (400 g) in females. The primary tissue sites of phosphorus storage are in bone-hydroxyapatite (85%) and skeletal muscles (14%).

HOMEOSTATIC REGULATION

Plasma phosphorus reflects the net rate of phosphorus flux between intestine, kidney, bone and soft tissue. The kidney is the primary regulator of the plasma phosphorus level by altering the rate of phosphorus reabsorption of the filtered phosphorus load. At low renal phosphorus loads essentially all of the filtered phosphorus is reabsorbed and none is excreted in the urine. As the amount of phosphorus in the glomerular filtrate increases, the capacity of the renal reabsorptive apparatus is exceeded and phosphorus appears in urine.

Regulation of total body phosphorus over long periods requires the coordinated effects of kidney and intestine. Under conditions of low dietary phosphorus intake the intestine increases its absorption efficiency and the kidney increases renal phosphorus reabsorption to minimize phosphorus loss. This is controlled by 1, 25 DHCC and PTH.

The effect of other minerals on phosphorus absorption include the following.

1. Phosphorus decreases lead absorption in humans.

2. Increased calcium absorption decreases phosphorus absorption.

3. Aluminium- or magnesium-containing antacids bind with phosphorus in GI tract and reduce phosphorus absorption.

FUNCTIONS AND MECHANISM OF ACTION

1. The major building blocks are covalent molecules, polymers, proteins, polysaccharides and nucleic acids. DNA and RNA are polymers based on phosphate ester monomers.

2. The high energy PO_4 bond of ATP is the major energy currency of cells.

3. Cell membranes are composed largely of phospholipids.

4. The inorganic constituents of bone are primarily a calcium phosphate salt.

5. A variety of enzymatic activities are controlled by alternate phosphorylation and dephosphorylation of proteins by cellular kinases and phosphatases.

6. The metabolism of all major metabolic substrates depends on the functioning of phosphorus as a cofactor in a variety of enzymes.

7. It is the principal reservoir for metabolic energy in the form of ATP, creatine phosphate and phosphoenol pyruvate.

8. Phosphates are ionized at physiological pH so that they may trap phosphorylated molecules within cells.

9. Phosphorus is also part of many coenzymes like TPP, NADP, FNIN, FAD, etc.

10. The combination of phosphorus (as PO_4) makes it possible for many nutrients to cross a cell membrane, be carried in the blood or to become a large enough molecule so that they do not "leak" out of a cell (phosphorylation).

11. It helps in the regulation of acid balance as Na_2HPO_4 and NaH_2PO_4.

DIETARY CONSIDERATIONS

Food sources In general, food sources high in protein (meats, milk, eggs and cereals) are also high in phosphorus.

Phosphorus from meat is well absorbed. It is found mainly as intracellular organic compounds that are mostly hydrolysed in the GI tract, releasing inorganic phosphorus that is available for absorption.

Phosphorus deficiency   Chronic phosphorus deficiency in animals results in the loss of appetite, development of stiff joints, fragile bones, and a marked increase in susceptibility to infection.

In normal individuals, phosphorus deficiency is believed to be unlikely to occur because of the widespread availability of phosphorus in the diet. However, premature infants are an exceptional case, because they are frequently prone to the development of rickets due to an inadequate supply of phosphorus and calcium. Severe phosphorus deficiency and depletion along with clinical symptoms are summarized in Table 9.1.

Table 9.1   Clinical manifestation of severe hypophosphataemia and phosphorus depletion

Neurologic	Parasthesia, confusion, seizures, coma, intention tremor, ataxia
Haematologic	Reduced red cell, haemolytic anaemia, impaired leucocytic chemotaxis, phagocytosis and bacterial killing, platelet dysfunction
Respiratory	Acute respiratory failure, hyper or hypoventilation
Muscular	Generalized weakness, proximal myopathy
Renal	Hypophosphaturia, hypercalciuria, hypermagnesuria
Cardiac	Reversible cardiomyopathy, depressed vascular response to vasopressors
Skeletal	Osteomalacia with bone pain and pseudofractures
Endocrine	Functional hypoparathyroidism, glucose intolerance

PHOSPHORUS REQUIREMENTS

RDA of phosphorus = RDA of calcium

Age group	Requirement
Birth to 6 months old infant	300 mg
Infants 6–12 months old	500 mg
Children 1–10 years old	800 mg
11–24 years old	1200 mg
24 years and above	800 mg

PHOSPHORUS EXCESS

When there is an excess of phosphorus in the body, hypocalcaemia, secondary hyperparathyroidism with excessive resorption and bone loss are seen. In infants it can cause hypocalcaemia and tetany.

REVIEW QUESTIONS

1. Explain the functions of calcium.
2. Discuss the factors affecting calcium absorption.
3. Explain the role of hormones in maintaining calcium levels.
4. Discuss the interrelationship of calcium, phosphorus and vitamin D.
5. Explain the deficiency of calcium.

CRITICAL THINKING QUESTIONS

1. Does phosphate help in maintaining acid–base balance?

2. Are major minerals more important to health than trace minerals?

3. Which is the best source of minerals—plant foods or animal foods?

4. Can calcium supplements prevent osteoporosis?

10

MICRO MINERALS

Did You Know?

- Trace mineral deficiencies are difficult to detect. Laboratory techniques are not even available to adequately measure some trace elements in tissues, and there are gaps in our knowledge about metabolism and storage of many minerals. This hampers interpretation of laboratory results.

- Trace minerals that have the same charge and are chemically similar or use the same carrier proteins often compete with each other during absorption and metabolism. This must be considered when setting the RDA for a trace mineral and when using mineral supplements.

- Trace mineral concentrations in plants usually depend on the trace mineral concentrations in the soil in which the plants are grown. Eating a variety of foods will ensure an intake from a variety of soil conditions and help maximize the chances of consuming an adequate amount of trace minerals.

- Enriched grains contain less of many trace minerals than whole grains. The refinement process strips a food, such as flour, of many minerals. The enrichment process only replaces one: iron.

- Iron is the only nutrient with a higher RDA for adult women than for adult men. This is because women in their childbearing years have a greater need for iron due to their greater iron loss during menstruation.

- The main function of zinc is promoting growth and development. A zinc deficiency can cause poor growth in children, whereas a zinc supplement can help increase appetite and the rate of weight gain in children.

- Taking high doses of zinc supplements often disrupts the natural nutrient balance found in a good diet and and affects immune function resulting in more harm than good.

IRON

Iron was used in very early years as a medicine. In the Ebers papyrus of Egypt, rust was prescribed as an ointment to prevent baldness. In early Greece, a solution of iron in wine was considered as a means to restore male potency. In the 17th century the treatment of chlorosis included iron supplementation. Later it was found that chlorosis is a iron deficiency.

Even before the role was in nutrition of iron firmly established, the first clinical description of iron overload disease was reported in 1871. The major aspects of iron metabolism were elucidated before 1960.

BIOLOGICAL IMPORTANCE

Iron is one of the most useful both in technology and in biology because iron compounds are involved in numerous oxidation–reduction reactions. Aerobic metabolism depends on iron as an electron carrier in cytochromes and as a means of O_2 and CO_2 transport in haemoglobin.

FUNCTION

Carrier of Oxygen and Carbon Dioxide

The major role of iron is to permit the transfer of oxygen and carbon dioxide from one tissue to another. It does this as a part of haemoglobin in blood and myoglobin in muscle but also as a part of several tissue enzymes essential in cell respiration. The exchanges are involved primarily in the release of energy within the cell.

Blood Formation

Haemoglobin is an essential component of the mature red blood cells. Erythrocytes begin in the bone marrow as immature cell known as erythroblasts, which contain a nucleus. As these cells mature in the bone marrow, they synthesize haem, an iron-containing protein from the amino acid glycine and iron along with vitamin B_6 and copper. Haem now unites with a protein called globin, and forms haemoglobin.

The RBC has no nucleus and so cannot synthesize the enzymes essential for metabolism. As a result it lives only as long as the enzymes present at its maturity remain functional—usually about 4 months. As RBCs die, they are removed from the blood by the cells of the liver, bone marrow and spleen. In the spleen, iron and amino acids of the haemoglobin molecule are salvaged and recycled.

Anti-infective Agent

During iron deficiency, there is a decrease in the production of iron-containing enzymes and other immune substances needed to destroy infectious organisms, and iron-deficient people are less able to fight infections. Lactoferrin in human milk is an iron-containing substance that is effective against *E. coli* organisms in the gastrointestinal tract of infants. Here lactoferrin binds iron and so it is not available for bacterial growth.

Other Functions

Iron performs many other important functions including the following:

- catalyses the conversion of β-carotene to vitamin A.

- It is involved in the synthesis of purines which are required for formation of DNA and RNA.

- It is involved in the synthesis of carnitine, which is needed for fatty acid transport.

- Synthesis of collagen, the structural protein surrounding the cell.

- It is necessary for antibody production.

◌ It is involved in the detoxification of drugs in the liver.

◌ It functions as a transport medium for electrons within cells.

◌ Some iron-containing enzymes function as signal-controlling substances in some neurotransmitter systems in the brain, for example, the dopamine and serotonin systems.

◌ Other key functions for the iron-containing enzymes (cytochrome P450) is the synthesis of steroid hormones and bile acids.

IRON METABOLISM

The metabolism of iron can be described as two loops, one internal and one external. The internal loop is represented mainly by the formation and destruction of red cells. When a red cell dies, after about 120 days, it is usually taken care of by the macrophages of the reticuloendothelial system of the body. The iron is released and delivered to the transferrin molecules in the plasma. Transferrin, a protein specially designated for iron transport in plasma, brings the iron back to the red-cell precursors in the bone marrow, or to other cells in different tissues under growth and development. There is thus a continuous reutilization of the iron from haemoglobin in the red cells back to new red cells or other tissues.

IRON BALANCE

Iron balance means that a steady state is present, when the absorption of iron from the diet covers both the actual losses of iron from the body and the requirements for growth and pregnancy.

The absorption of non-haem iron is markedly influenced by the iron stores. The more the iron store the less iron is absorbed. An increased erythropoiesis is seen after an acute blood loss which also increases iron absorption.

The amount of iron lost with menstruation is influenced by the haemoglobin level. During the development of an iron-deficiency anaemia, menstrual iron losses will thus successively decrease when the haemoglobin level diminishes. Skin iron losses will also decrease in a state of iron deficiency and increase if a subject develops iron overload. Sooner or later, a point is reached when iron losses are equal to the amount of iron absorbed from the diet.

Iron absorption is higher in women with anaemia and lower in woman with iron stores. Hence, iron status influences the "setting" of the ability of the mucosal cells to absorb iron. Thus the regulation of iron absorption in relation to iron status tries to maintain balance.

Mechanisms for Maintenance of Iron Balance

The body has three mechanisms for maintaining iron balance and preventing iron deficiency.

1. The continuous reutilization of iron from catabolized red cells in the body.

2. The regulation of the absorption of iron from the intestines, with an increased iron absorption in the presence of iron deficiency and a decreased absorption in states of iron overload.

3. The access of the specific storage protein ferritin, which can store and release iron to meet excessive iron demands, such as in the last trimester of pregnancy.

Compartments

In humans the total quantity of body iron varies with weight, haemoglobin concentration, sex and size of the storage compartment.

Table 10.1 Iron compartments in normal man

	75-kg Man (mg)	55-kg Woman (mg)
Functional iron		
Haemoglobin	2400	1600
Myoglobin	350	230
Haem enzymes	150	110
Non-haem enzymes	3	2
Transferrin-bound iron		
Total functional iron	~ 2900	~ 1940
Storage iron	500–1500	0– 300
Ferritin and haemosiderin		
Total	~ 4000	~ 2100

Of these, the largest compartment is iron in haemoglobin, contained within circulating erythrocytes. The size of this compartment varies considerably according to body weight, sex and blood haemoglobin concentration (Table 10.1).

ABSORPTION

In healthy persons who do not lose iron by bleeding, iron loss is limited. Therefore, normal iron balance is maintained largely by regulation of iron absorption. Ingested inorganic iron is solubilized and ionized by the acid gastric juice, reduced to the ferrous (Fe^{2++}) form and chelated.

Substances that form low molecular iron chelates, such as ascorbic acid, sugars and amino acids, promote iron absorption.

There is impaired iron absorption in achlorhydria and gastrectomized subjects due to decreased solubilization and chelation of ferric iron in food.

Absorption may occur at any level of the small intestine, but it is most efficient in the duodenum. The mucus itself has iron-binding properties. The passage of iron through the mucus is facilitated by organic acids or taurocholic acid in bile or by polypeptides derived from meat, fish or poultry.

Fe^{2+} traverses the mucous layer more readily to reach the brush border of the intestinal epithelial cells. There Fe^{2+} is oxidized to Fe^{3+} before it enters the epithelial cells.

At the cell membrane of the brush border of the epithelial cell, Fe^{3+} is bound to a receptor protein called membrane iron-binding protein (MIBP) which then transfers iron into the cell. Apotransferrin in the cytosol of intestinal epithelial cells may accelerate iron absorption.

Most of the iron that is absorbed from the intestinal lumen passes rapidly through the mucosal cells in the form of small molecules. As the iron enters, the plasma it is oxidized to Fe^{3+} by ceruloplasmin which functions as a ferroxidase, and then it is taken up by transferrin.

Excess of Fe^{3+} combines with apoferritin to form ferritin. Some may later be released into the circulation, but most of it remains in the mucosal cells until they are sloughed into the intestinal lumen at the end of their 2–3-day lifespan.

Haem iron is absorbed directly after removal of the globin by proteolytic enzymes. Once inside the mucosal cell, the iron is liberated from haem probably by the enzyme haem-oxygenase.

TRANSPORT

Iron is transported by a protein called transferrin. The normal concentration of transferrin in plasma is about 2.2 to 3.5 g/L (220 to 352 mg/dl). Because iron is the natural ligand of transferrin, the plasma concentration of transferrin may be measured by the amount of iron it will bind. This is called the total iron-binding capacity (TIBC). One atom of Fe^{3+} can be bound together with a bicarbonate ion, at each end of the transferrin molecule (diferric transferrin).

Cell membranes contain a protein called transferrin receptor. Early erythrocyte precursors have abundant transferrin receptors on their membranes.

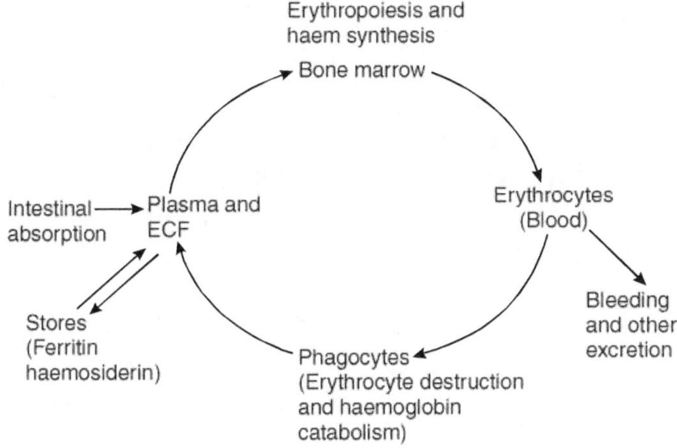

Figure 10.1   Pathway of iron metabolism

Iron is tightly conserved, and each iron atom cycles repeatedly from plasma and ECF to the bone marrow where it is incorporated into the erythrocytes and circulates for 4 months (Figure 10.1).

ROLE OF COPPER IN IRON METABOLISM

Copper is essential for utilization of iron for haemoglobin formation. Two ferroxidases containing copper (ceruloplasmin and ferroxidase II) are involved in the conversion of Fe^{2+} to Fe^{3+} in the intestinal mucosa and in plasma. The oxidation of ferrous iron to ferric form and vice versa is necessary for movement of iron from intestinal lumen to the mucosa, blood and bone marrow for incorporation in haem.

STORAGE IRON

Iron in excess of need is stored intracellularly as ferritin and haemosiderin principally in the macrophage (reticuloendothelial) system of liver, spleen, bone marrow and other organs. Ferritin is the basic storage molecule for iron; haemosiderin appears to be aggregated ferritin partially stripped of its protein component.

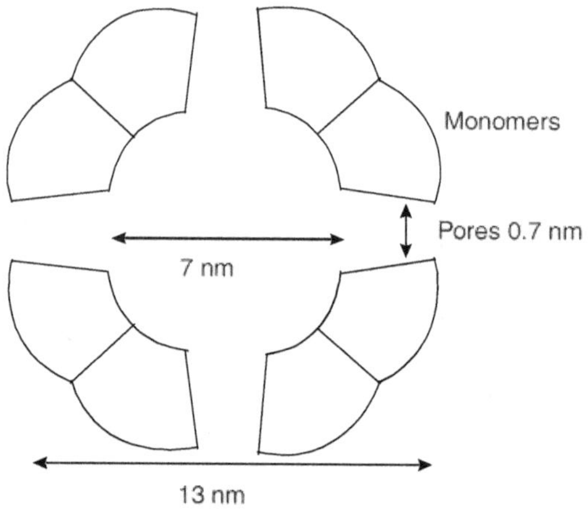

Figure 10.2   Ferritin molecule

A complete ferritin molecule consists of an apoferritin protein shell that is 13 nm in diameter and has an internal cavity of 7 nm in diameter (Figure 10.2). Within the internal cavity are one or more crystals of ferric oxyhydroxide, FeOOH, together with trace amounts of phosphate. The cavity of each ferritin molecule has the capacity to hold at maximum 4300 iron atoms. The protein shell has pores which are approximately 0.7 nm in diameter, just large enough to permit monosaccharides, flavin mononucleotide, ascorbic acid and desferrioxamine to enter the interior cavity. Furthermore, the pores appear to function as catalytic sites for the binding of Fe^{2+}, its oxidation to FeOOH and the passage of FeOOH to the growing core crystal. Thus apoferritin shell is not only an efficient iron trap but also functions enzymically (Figure 10.3).

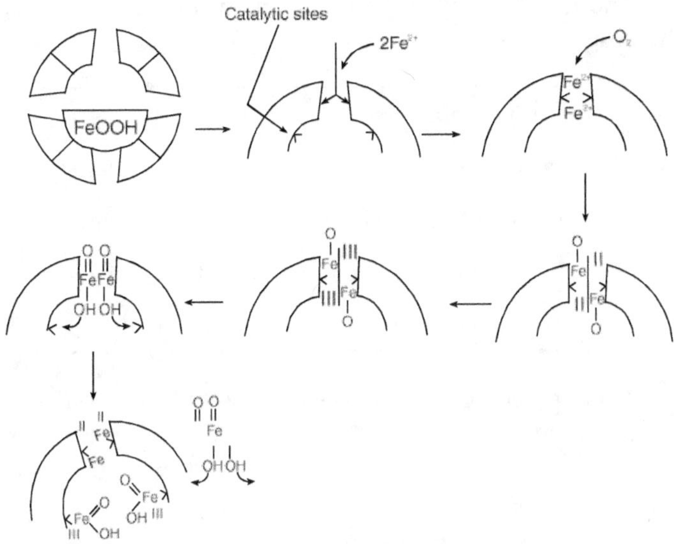

Figure 10.3   Scheme of uptake and deposition of iron by ferritin

Iron release from ferritin is mediated by reduced FMN.

Iron Loss

The body has a limited capacity to excrete iron. Daily iron loss in adult men is between 0.9 and 1.05 mg or approximately 0.013 mg/kg of body weight.

Menstruation

The mean menstrual loss is 43 ml equivalent to an average of about 0.6 to 0.7 mg of iron per day. Iron loss is as follows:

GI blood (haemoglobin)	0.35 mg
GI mucosa (ferritin)	0.1 mg
Biliary	0.2 mg
Urinary	0.08 mg
Skin	0.2 mg

The iron cost of pregnancy is high. In a normal pregnancy, the cost of iron is as follows:

Iron contributed to foetus	200–370 mg
In placenta and cord	30–170 mg
In blood loss at delivery	90–310 mg
In milk, lactation 6 months	100–180 mg
Total	420–1030 mg

Average/day (pregnancy—9 months and lactation—6 months) 1–2.5 mg/day

FACTORS AFFECTING DIETARY IRON ABSORPTION

Haem iron absorption

- Amount of haem iron present in meal
- Content of calcium in meal
- Food preparation time and temperature

Non-haem iron absorption

- Iron status of subjects
- Amount of bioavailable non-haem iron
- Balance between dietary factors enhancing and inhibiting iron absorption

Factors enhancing iron absorption

- Ascorbic acid
- Meat, fish, seafood
- Certain organic acids

Factors inhibiting iron absorption

- Phytates
- Iron-binding phenolic compounds
- Calcium
- Soy protein

IRON OVERLOAD

The amount of storage iron in the body seldom exceeds 2000 mg. In fact, in women, it seldom exceeds 400–500 mg and in men 1500 mg. Increased amounts may occur by increased parenteral administration of iron given as parenteral iron preparations or as blood transfusions or by increased absorption from the intestines.

A higher intake of dietary iron will lead to increased absorption. However, as the amount of iron in store increases, the fraction of dietary iron absorbed will successively decrease. Moreover, the losses of iron from the body will increase due to the higher iron content of the desquamated cells. Therefore, in otherwise healthy subjects, a high dietary intake of iron will not lead to any organ damage due to a pathological accumulation of iron. The only "nutritional" iron overload condition described in healthy people is in the Bantu population consuming large amounts of a special beer with a very high iron content, brewed in iron containers. Iron overload has also been observed in alcoholics consuming large amounts of wine with a high iron content.

Iron overload may also occur in patients who have certain chronic anaemias with an increased, ineffective erythropoiesis which induces an increased absorption of dietary iron as in thalassemia major.

In idiopathic hereditary haemochromatosis, the absorption of dietary iron is increased, resulting in serious organ damage. For example, in the liver the iron overload may cause cirrhosis, in the pancreas diabetes, in the testis hypogonadism and in the heart serious cardiac disturbances. The iron overload should be diagnosed and corrected early before organ damage becomes clinically manifest.

Definition of Some Important Terms

Siderosis   Presence of excess iron in the tissues. Iron stored in this form is not available to the body.

Nutritional siderosis   Large amounts of iron in liver is seen in the Bantu tribe of South Africa. Haemochromatosis a rare disease in which excessive amounts of iron are absorbed from the intestinal tract. The absorbed iron is deposited chiefly in the liver, spleen, pancreas, skin and several other organs.

IRON DEFICIENCY

When iron can no longer be mobilized from the iron stores, insufficient amounts of iron will be delivered to transferrin, the circulating transport protein for iron. The binding sites for iron on transferrin will thus contain less and less of iron. This is called reduction of the transferrin saturation (TS). When the transferrin saturation drops to about 16% which is considered to be a critical level, the red-cell precursors get an insufficient supply of iron. Simultaneously the supply of iron to other tissues by transferrin will also be impaired. Cells with a high turnover rate, e.g. intestinal mucosa cells with a short lifespan are immediately affected.

Hence an iron deficiency can be seen in growing tissues.

The haemoglobin formation is also impaired. When the deficiency is sufficiently severe and longstanding will lead to a microcytic hypochromic anaemia.

Diagnosis

i. Measurement of stainable iron in the bone-marrow smears.

ii. A low concentration of ferritin in the serum about 15 μg/l or less.

iii. Low transferrin saturation and a high content of transferrin receptor.

Causes of Iron Deficiency

Nutritional iron deficiency implies that the diet cannot cover physiological iron requirements worldwide. This is the most common cause of iron deficiency.

The pathological causes of iron deficiency include the following.

- In many tropical countries, infestation with hookworm leads to intestinal blood loss which may be considerable.

- Pathological losses of blood could be because of tumours in the GI tract or uterus.

- Achlorhydria occurs when the patient absorbs less iron.

- Patients who have undergone gastric surgery are deficient in iron.

- A largely vegetarian diet can be a factor in causing anaemia.

Deleterious Effects of Iron Deficiency

Negative effects of iron deficiency on important functions of the body.

- Physical working capacity is reduced due to the impaired oxidative metabolism in the muscles, i.e., a reduction in the iron-containing enzymes which are rate-limiting for oxidative metabolism.

- Several structures in the brain have a high iron content of the same magnitude found in the liver. There is an active transferrin-receptor-mediated transport of iron into the brain. The supply of iron to brain cells mainly takes place during an early phase of their development, and thus an early iron deficiency may lead to irreparable damage.

- There is an impairment in memory and learning, an increase in the pain threshold, a reduction in the release of thyrotropin release hormone and hence a reduction in the thyroid function and thermoregulation in the body.

- Iron deficiency also impairs the activity of normal defence systems against infection. The cell-mediated immunological response by the action of T lymphocytes is impaired due to a reduced formation of these cells, due to a reduced DNA synthesis depending on the function of ribonucleotide reductase, which requires a continuous supply of iron for its function.

- The phagocytosis and killing of bacteria by the neutrophil leucocytes is also affected in iron deficiency due to absence of iron containing enzymes.

- A reduction in physical working capacity has been seen in population with long-standing iron deficiency.

- A relationship between iron deficiency and behaviour such as attention, memory and learning has been demonstrated in infants and small children.

- Mostly tongue, nails, mouth and stomach are affected. The skin appears pale and the inside of the lower eyelid may be light pink instead of red. Finger nails can become thin and flat and eventually koilonychia (spoon-shaped nails) develops. Mouth changes include atropy of the lingual papillae, burning redness and in severe cases a completely smooth waxy and glistening appearance to the tongue (glossitis).

- Angular stomatitis and dysphagia may occur. Gastritis occurs frequently and may result in achlorhydria. Progressive, untreated anaemia results in cardiovascular and respiratory changes that can eventually lead to cardiac failure.

Other symptoms include lassitude, fatigue, breathlessness on exertion, palpitations, dizziness, tinnitus, headache, dimness of vision, insomnia, paraesthesia in fingers and toes and angina. Growth abnormalities are also seen in anaemic children.

Treatment

Oral iron therapy is usually very effective if well absorbed iron tablets are given at a sufficient dosage for a sufficiently long time. A dosage of about 50–200 mg (60 mg of elemental iron) is given as ferrous sulphate.

Iron therapy should be continued for several months even after restoration of normal haemoglobin levels, to allow for repletion of body iron stores.

In iron deficiency anaemia when dietary folate is also low, treatment with iron alone precipitates folate deficiency because more is needed for production of erythrocytes and the supply of folate becomes insufficient.

Prevention

This is achieved by selecting one or more of the following strategies.

1. Dietary improvement which advocates a regular consumption of iron-rich foods.

2. Iron supplementation, i.e., giving iron tables to certain target groups such as pregnant and preschool children.

3. Iron fortification of certain foods.

4. Food education

    - Educating people about the rich sources of iron and increasing its bioavailability

    - Increase an awareness about the symptoms of anaemia

    - Promotion of home gardening

IODINE

Iodine is an essential constituent of the thyroid hormone thyroxine. The major role of iodine in nutrition arises from the importance of thyroid hormones to growth and development.

Iodine was discovered by Courtois in 1811 during the course of making gunpowder. Some seaweed ash was being used from which the iodine vaporized as a violet vapour. The element was discovered in the thyroid gland by Baumann in 1895.

The relation of iodine deficiency to the enlargement of the thyroid gland or goitre was first shown by David Marine, who found that hyperplastic changes occurred regularly in the thyroid when iodine concentration fell below 0.1%. Subsequently Marine and Kimball (1922) demonstrated that endemic goitre could be prevented and substantially reduced by the administration of small amounts of iodine.

Mass prophylaxis of goitre with iodized salt was first introduced in Switzerland and in Michigan.

PHYSIOLOGY

The healthy human adult body contains 15 to 20 μg of iodine of which about 70 to 80% is in the thyroid gland. The thyroid gland, which weighs only 15 to 25 g possesses a remarkable concentrating power for iodine. The amount of iodine in the gland is closely related to the iodine intake.

Iodide is rapidly absorbed through the gut. The normal intake and requirement is 100 to 150 μg per day. Excess iodine is excreted by the kidney. The level of excretion correlates with the level of intake, so that it can be used to assess the level of iodine intake.

Iodine exists in the thyroid as inorganic iodine; the iodine-containing amino acids, monoiodothyronine (MIT), diiodothyronine (DIT), T3 and T4; polypeptides containing thyroxine and thyroglobulin. Thyroglobulin is a protein that contains the iodinated amino acids in a peptide linkage.

Thyroglobulin is the chief constituent of the colloid that fills the thyroid follicle. It is a glycoprotein (it contains glucose). It is the storage form of the thyroid hormones and makes up 90% of the total iodine in the gland.

Iodine occurs in food either as reduced iodide, as inorganic iodine or bound to an organic component from which it should be freed. Free iodine is reduced to iodide and absorbed primarily in the small intestine.

Absorbed iodide appears immediately in the bloodstream from which one-third is absorbed by the thyroid gland and the rest is taken up by the kidneys to be excreted in the urine, a protection against the accumulation of toxic levels. Small amounts are lost in perspiration and excreted in the faeces.

The thyroid has to trap about 60 µg of iodine per day to maintain an adequate supply of thyroxine. This is possible because of the very active iodide trapping mechanism, which maintains a gradient of 100 : 1 between the thyroid cell and the extracellular fluid. In iodine deficiency, this gradient may exceed 400 : 1 in order to maintain the output of thyroxine.

The trapping of iodide by the thyroid depends on an active transport mechanism (i.e., it requires energy) which is called as the "iodine pump". This mechanism is regulated by the thyroid stimulating hormone (TSH) which is released from the pituitary to regulate thyroid secretion.

Other ions can compete with iodide, including thiocyanate (SCN).

The iodide is released by the thyroid cells into the colloid follicle phase between the cells, and there it is oxidized by hydrogen peroxide from the thyroid peroxidase system. It then combines with tyrosine in the thyroglobulin to form MIT and DIT. The oxidation process then continues with the coupling of MIT and DIT to from iodotyrosines (Figure 10.4).

Finally the iodinated thyroglobulin, including the iodinated amino acids, is absorbed back into the thyroid cells by pinocytosis.

It is then exposed to proteolytic enzyme, which breaks it down to release T4 and T3 into the blood.

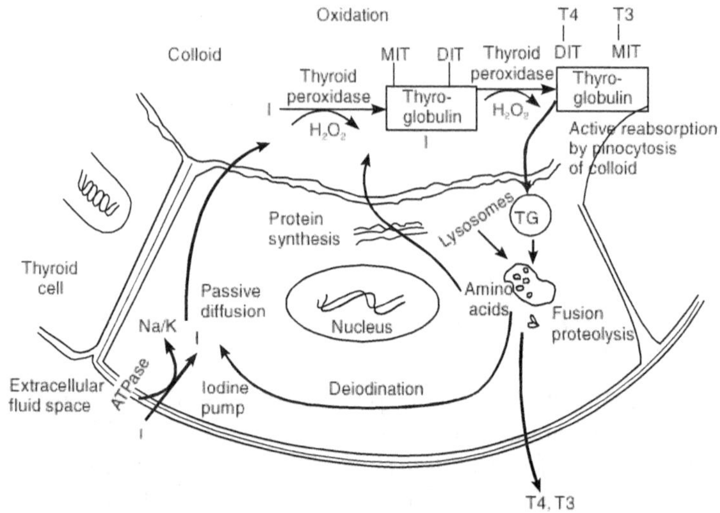

Figure 10.4   Synthesis of T3 and T4

The unused iodotyrosines are conserved for incorporation into a subsequent cycle of the biosynthetic process.

The regulation of thyroid hormones is a complex process involving not only the thyroid, but also the pituitary, the brain and the peripheral tissues.

The thyroid secretion is under the control of the pituitary gland through TSH.

DEVELOPMENT OF GOITRE

The process of production and regulation of thyroid hormones provides the framework for understanding the production of goitre as a result of iodine deficiency. Although not the only cause, iodine deficiency is the primary cause of goitre. Goitrogens such as thiocyanates that can enhance the effect of iodine deficiency are referred to as secondary factors.

The basic effect of iodine deficiency is to interfere with the production of thyroid hormones because iodine is an essential constituent of the T4 and T3 molecules. The lowering of output from the thyroid leads to a fall in the blood levels of T4 but to some increase in T3 (the less iodinated hormone is produced preferentially in iodine deficiency).

The fall in the level of T4 leads to increase in TSH output from the pituitary. TSH increases the uptake of iodide by the thyroid with increased turnover associated with hyperplasia of the cells of the thyroid follicles. The reserves of colloid containing thyroglobulin are gradually used up so that the gland has a much more cellular appearance than normal. The size of the gland increases with the formation of a goitre (Table 10.2).

IODINE DEFICIENCY DISORDERS

The effects of deficiency are evident at all stages, the foetal stage, neonatal stage and infancy which are periods of rapid growth (Table 10.2).

Table 10.2   The spectrum of iodine deficiency disorders (IDD)

Foetus	Abortions, stillbirths, congenital anomalies, increased perinatal mortality, increased infant mortality, neurologic cretinism (mental deficiency, deaf mutism, spastic diplegia, squint).
	Myxedematous cretinism (dwarfism, mental deficiency) psychomotor defects.
Neonate	Neonatal goitre, neonatal hypothyroidism.
Child and adolescent	Goitre, juvenile hypothyroidism, impaired mental function, retarded physical development.
Adult	Goitre with its complications, hypothyroidism, impaired mental function.

DEFINITIONS OF GOITRE STAGES

Goitre A normal thyroid gland should have the minimal size compatible with euthyroidism under conditions of normal iodine intake (100 to 150 µg/day). This gland would be non-palpable or barely palpable. For practical purposes a thyroid gland whose lateral lobes have a volume greater than the terminal phalanges of the thumbs of the person examined is defined as goitrous.

Estimation of Thyroid Size

Stage 0   No goitre

Stage 1a   Goitre detectable

Stage 1b   Goitre palpable and visible only when neck is fully extended

Stage 2   Goitre visible when neck is in normal position

Stage 3   Very large goitre easily recognized

An area is arbitrarily defined as endemic with respect to goitre if more than 10% of the population or of the children aged 6 to 12 years are found to be goitrous.

IODINE-INDUCED HYPERTHYROIDISM

A mild increase in incidence of hyperthyroidism has now been identified following iodized salt programs. This condition is seen in those over 40 years of age.

CORRECTION OF IODINE DEFICIENCY

Iodized salt In India the iodization of salt has been made mandatory. The cost of such iodized salt is small in relation to the advantages. However there is still the problem of the salt actually reaching the iodine-deficient subject. There may also be a problem with the distribution or preservation of the iodine content, i.e., the salt may be left uncovered or left exposed to heat. It should be added after cooking to reduce loss of iodine.

Iodized oil by injection In some countries like Zaire, South America and China the correction of severe deficiency is by a single intramuscular injection (1 ml of oil contains 480 mg).

Iodized oil by mouth In China, oral iodized oils are being given.

ASSESSMENT OF IODINE STATES

1. **The goitre rate** The incidence or rate of goitre in the population.

2. **Urine iodine excretion** Excretion of iodine is a good index of the level of iodine nutrition.

3. Serum levels of T4 or TSH is a good measure of iodine status.

IODINE REQUIREMENTS

NRC (National Research Council) has recommended a daily intake of 40 mg for children aged 0–6 months, 50 µg from 6 to 12 months, 70 to 120 µg from 1 to 10 years and 120 to 150 µg from 11 years onwards. The recommended intake during pregnancy and lactation were respectively 175 and 200 µg.

IODINE TOXICITY

A daily intake of more than 2000 µg is considered as potentially harmful.

Normal diets composed of natural foods are unlikely to supply such a high value of iodine except where the diets are exceptionally high in marine fish or seaweed or where foods are contaminated with iodine from adventitious sources.

It has been shown that

1. Normal subjects who have not been iodine-deficient can maintain normal thyroid function states even when they are taking several milligrams of dietary iodine per day.

2. Incidence of non-toxic diffuse goitre and toxic nodular goitre are remarkably decreased by high dietary iodine.

3. Incidence of Grave's disease and Hashimoto's disease appears not to be affected by high dietary iodine.

4. However, high dietary iodine may induce hypothyroidism in autoimmune thyroid diseases.

SOURCES

Drinking water in goitrous areas is 3–16 µg/l as compared to 5–6 µg/l in other areas.

COPPER

The adult body contains about 100–180 mg of copper. It is present in all tissues but liver, brain, heart and kidney contain the highest amounts.

In blood, copper is present in equal amounts in plasma (0.115 µg/100 ml) and erythrocytes (0.11 µg/100 g).

About 90% of plasm a copper is present in ceruloplasmin while 60% of RBC copper is present in a protein called erythrocuprein.

0.3% copper is present in a protein called cerebrocuprein which is present in the brain. Adult liver contains about 10–15 mg copper, kidney contains 10 mg and brain also contain 10 mg.

FUNCTIONS OF COPPER

1. Copper plays a part in preventing anaemia by

   i. aiding in iron absorption

   ii. stimulating the synthesis of haem or globin

   iii. releasing stored Fe from ferritin

   iv. ferroxidase I and II or ceruloplasmin are involved in the oxidation of ferrous to ferric ion.

2. Required for synthesis of phospholipids which are essential in the formation of myelin surrounding nerve fibres.

3. It is a part of the respiratory enzyme cytochrome oxidase which is necessary in electron transport chain for the release of energy in the cell.

4. In conjunction with vitamin C, it maintains the activity of the enzyme Lysl oxidase which is involved in the synthesis of both elastin a protein in the wall of the aorta and collagen which is a part of the connective tissue.

5. It is a part of the enzyme tyrosinase needed for the conversion of the amino acid tyrosine to melanin which is the dark pigment of hair and skin. The absence of this enzyme is associated with albinism, a condition in which there is no colour in the hair or eyes.

ABSORPTION AND METABOLISM

Typical diets provide 1 mg or less of copper out of which 2–10% is absorbed. Copper is taken up by the stomach and upper intestine where the contents are still acidic. Its absorption from the intestine is dependent on a copper-binding protein called metallothionein. Absorption of copper decreases if intake of vitamin C is high.

Absorbed copper appears in the bloodstream after 15 minutes of being consumed. Initially it is loosely bound to albumin or to some amino acid to produce chelates (combination of trace element and another substance). These chelates aid in the transport of copper across the membranes and into the cells.

Transport of copper represents only 7% serum copper.

Copper is removed from the blood serum by the liver. From there copper is either,

i. excreted into the bile.

ii. stored in a protein complex containing 2% copper.

iii. used in the synthesis of ceruloplasmin which is a protein–copper complex that is released into the blood and accounts for 60% of serum copper.

Serum copper values are not a valid measure of copper status because plasma copper can be maintained at the expense of liver stores.

The release of copper from the liver is controlled by the adrenal gland. Some serum copper enters the bone marrow where it is used in the synthesis of erythrocuprein.

Erythrocuprein is identical to hepatocuprein in the liver and cerebrocuprein in the brain.

All these three compounds are referred to as cytocuprein or superoxide dismutase (SOD).

EXCRETION

Copper is excreted both in faeces and urine. Faecal copper includes unabsorbed dietary copper, copper released in the bile and copper lost through intestinal wall. Urinary copper accounts for 4% of copper loss.

BLOOD COPPER LEVELS

The copper content of plasma is 0.115 µg/dl and that of erythrocytes is 0.11 µg/100g. The level is not changed appreciably as a result of ingestion of copper or during fasting. The plasma levels increase during pregnancy.

Low plasma copper levels are found in individuals suffering from Wilson's disease. In Wilson's disease, ceruloplasmin content of plasma is also reduced from 34 mg/100 ml to 9 mg/100 ml.

RDA OF COPPER

Infants	0–1 year	0.5–1 mg/kg body weight (premature infants need higher amounts)
	1–3 years	1.0–1.5 mg/day
	4–6 years	1.5–2 mg/day
	6 years–adult	2–3 mg/day

Pregnant women need an additional 0.3 mg/day as 20 mg has to be transferred to the foetus.

The copper content of some foods is given below.

Food	Copper content μg/100 g
Cereals	0.19–0.74
Pulses	0.66–1.25
Nuts and oilseeds	0.08–0.27
Roots and tubers	0.13–0.2
Green leafy vegetables	0.08–0.53
Other vegetables	0.05–0.23
Fruits	0.13–0.4
Milk and eggs	0.22–0.25
Meat and fish	0.22–0.27
Liver	2.8

GENERAL DEFICIENCY OF COPPER

Vegetarians have a higher risk of deficiency because the major food sources are non-vegetarian, and plant sources carry considerable dietary fibre which interferes with absorption.

Low copper plasma levels or hypocupraemia may be due to

1. Urinary losses of ceruloplasmin which has been observed in nephrosis.

2. Malabsorption diseases like celiac sprue because of decreased copper absorption.

Copper deficiency is rare in humans but was found in 7–9-month old infants on a milk diet who were hospitalized because of malnutrition associated with severe diarrhoea.

The deficiency is recognized by a drop in ceruloplasmin (copper-carrying protein) levels and low blood copper levels. Infants are born with stores of copper that normally last until food other than milk are introduced at 3–6 months.

The copper content of hair decreases with age and a condition known as Menke's kinky hair syndrome may develop which is characterized by

1. stubby white hair

2. slow growth

3. degeneration of brain tissue so mental retardation is seen

4. hypothermia

5. degeneration of skeletal and cardiac muscles and is associated with low serum copper and ceruloplasmin levels. This condition results from defective copper absorption when copper is taken up by intestinal cells and not released into the bloodstream.

COPPER TOXICITY OR WILSON'S DISEASE

It is a rare hereditary disorder of copper metabolism. In Wilson's disease about 50% of ingested copper is absorbed as compared to the normal 2–10%.

Copper is toxic when it exists as the unbound copper ion as it acts as an inhibitor to many enzyme systems. There is an accumulation of copper in the liver, brain and kidney and cornea of the eye where it is identified by brown or green rings.

If intake of copper salts is 10% higher than normal it can lead to nausea and vomiting.

Clinical Features of Wilson's Disease

The biochemical abnormality in Wilson's disease is present from birth but clinical signs do not appear until adolescence.

1. Hepatic type—liver cirrhosis eventually leading to liver failure.

2. Cerebral type—progressive dementia and loss of emotional control and Parkinsonism.

3. Eye lesions—Causes a golden brown yellow or green ring around the cornea. This lesion is called Kayser Fleischer ring.

Biochemical Disorders

1. Absorption of copper from intestine is high (50%).

2. Defect in the formation of ceruloplasmin.

3. Abnormal copper-binding protein synthesis in the liver, kidney and brain. In liver the protein called copper thionein binds 4 times more copper than ceruloplasmin.

4. Increased deposition of copper in liver, brain, etc. due to increased non-ceruloplasmic copper in plasma and high copper-binding proteins in these organs.

5. Urinary copper excretion is increased.

Treatment

Progressive and proves fatal. Copper binding drugs like disodium calcium versenate, and pencillamides are useful as they increase copper excretion and bring down stores to normal levels.

INDIAN CHILDHOOD CIRRHOSIS

It is a fatal disease of children aged 1–3 years which occurs in families mainly in the middle income range in rural areas.

There is an accumulation of copper in the liver which may be due to frequent usage of copper vessels especially for boiling and storing milk.

FLUORINE

The role of fluorine in controlling tooth decay has been recognized for some time. In addition, there is evidence that growth and possibly reproduction are dependent on a supply of fluorine, but there is lack of agreement on whether it is an essential nutrient.

In 1931 it was shown in some communities that people had a remarkable freedom from tooth decay but also suffered from dental fluorosis. In these communities there was a much higher level of fluorine in their water supplies. By 1942 it was established that water supplies containing 1 ppm of fluorine were associated with a 50% to 60% reduction in tooth decay.

METABOLISM

Fluoride is readily absorbed from the stomach, although some fluoride continues to be taken up by the intestine. Of the 90% of the fluoride that is absorbed by the body, half is excreted and half readily becomes an integral part of the tooth and bone structure. The kidneys control the level of fluoride in the blood by excreting the excess in the urine. The amount of fluorine appearing in the tissues, including soft tissues, saliva, milk and foetal blood, parallels the amount in blood but at a slightly lower level.

MODE OF ACTION

Where fluorine is available, some crystal, of fluorapatite replace the crystals of hydroxyapatite normally deposited during tooth formation. Fluorapatite in tooth enamel is less soluble in acid and more resistant to the cavity-producing action of acids.

Fluorine available in the saliva is absorbed on the tooth surface to add strength and rigidity. Crystals of fluorapatite are deposited on the bone also. As much as 5000 to 6000 ppm of fluorine in bone is physiologically safe.

Another benefit of fluorine intake includes greater stability of the skeleton to protect against the loss of calcium that occurs during menopause, when the body is immobilized and during space flights.

SOURCES

Drinking water has 1 ppm or more of fluorine. Tea provides 0.25 mg of fluoride/day and fish products especially mackerel and salmon are good sources (2.7 µg/100 g). Soybeans also contain about 0.4 to 0.67 µg/100 g and the use of fluoridated paste is also a source.

CLINICAL ABNORMALITIES

Acute toxicity which requires a daily intake of 2 to 10 g is rare. Chronic toxicity also occurs infrequently and is usually the result of prolonged use of water with a high natural fluorine content of 20 to 80 mg/day.

Dental Fluorosis

In many parts of the world where the drinking water contains excessive amounts of fluorine about 3.5 ppm, signs and symptoms of dental fluorosis are observed. The enamel loses its lustre, becomes rough, chalky white patches with a secondary infiltration of yellow or brown stains are found irregularly over the surface of the teeth. The enamel is structurally weak and in severe cases there is marked loss of enamel accompanied by "pitting" which gives the tooth surface a corroded appearance.

Skeletal Fluorosis

This is chronic fluorine intoxication through drinking water which contains excessive amount of fluorine over 10 ppm or among workers handling fluorine-containing minerals, and results in pathological changes in the bones.

There is sclerosis, increased density and hypercalcification of bones of spine, pelvis and limbs. In addition, the ligaments of the spine becomes calcified producing a "poker-back" stiff position. It is possible that the collagen in the bones is calcified. There may also be ossification of tendons, neurological disturbances, and secondary changes in the vertebral column are common.

Such individuals become crippled.

Prevention of Toxic Fluorosis

This can be prevented by removing the fluoride from water supply by treatment with inactivated carbon or by some other suitable absorbant.

Deficiency

Dental caries and osteoporosis may result.

REQUIREMENT

0–6 months	0.1–0.5 mg
6–12 months	0.2–1 mg
1–3 years	0.5–1.5 mg
4–10 years	1–2.5 mg
Adolescents	1.5–2.5 mg
Adults	1.5–4 mg

ZINC

Although zinc has been known to be a dietary essential in rats since 1934, it was not until 1961 that a zinc deficiency in humans was recognized. In 1974, with a knowledge of the metabolic role of zinc, the Food and Nutrition Board established an RDA for dietary zinc.

The human body contains from 2 to 2.5 g of zinc with three-fourths of this amount concentrated in the skeleton from which it is removed only very slowly. A high concentration of zinc appears in the skin, hair and testes. In the blood most of the zinc occurs in the red blood cells; a small amount is found in white blood cells, platelets and the blood serum. About a third of serum zinc is tightly bound to the protein macroglobulin, and the remainder is loosely bound to the protein albumin or to the amino acids histidine and cysteine. Blood serum levels of zinc fall during infections, pernicious anaemia, hyperthyroidism, pregnancy and the use of oral contraceptives.

ABSORPTION AND METABOLISM

Zinc is absorbed primarily in the upper part of the small intestine. Absorption is regulated in the cells of the intestinal wall which responds to zinc levels in the plasma. It is likely that zinc absorption is aided by a ligand or binding substance, such as dietary histidine citric acid, or one that is secreted in pancreatic juices. Zinc enters the mucosal cells (attached to the protein metallothien) lining the intestinal tract where it is either stored until the mature cells are sloughed off, utilized in the metabolism of the absorbed cell, or released to pass through to the blood. In the blood zinc is bound to a protein and carried to the liver, where it is stored until needed. One-third to half of ingested zinc is absorbed. Increased needs during pregnancy and lactation are reflected by up to a two-fold increase in absorption.

Absorption is decreased as the stores of zinc increases or when the diet contains large amount of whole grains in which either phytic acid or fibre may bind zinc as an insoluble complex. Oxalic acid has a similar effect (oxalic acid is found in certain vegetables). High levels of dietary calcium or copper also result in lowered absorption of zinc.

Non-haem iron interacts competitively with zinc and interferes with absorption when its ratio to zinc reaches 2 : 1 (because they compete for the common carrier, transferrin). Human milk contains a zinc-binding substance which enhances absorption.

Zinc is excreted in faeces. Unabsorbed zinc along with zinc from mucosal cells and that secreted in the pancreatic enzymes is excreted in the faeces. A small portion is lost in the urine (0.4 to 0.6 mg), in sweat (1 to 3 mg) and in semen (1 mg).

FUNCTIONS

1. It is a part of many metalloenzymes, e.g. carbonic anhydrase, carboxypeptidase, etc.

2. Zinc is important in the stabilization of membranes.

3. Zinc is necessasary in the protein-synthesizing structures in the cell.

4. Zinc is present as a free ion within the cell where it functions as a cofactor in many reactions.

5. **Reproduction** It plays a role in spermatogenesis and in the development of primary and secondary sex organs in males and all phases of the reproductive process in the female as zinc has the most profound influence on rapidly growing tissues.

    Zinc deficient men have shown retarded gonadal development. Impaired sexual maturity in both sexes is another result of zinc deficiency.

6. It is necessary for the development of Purkinje cells in the brain. Maternal zinc deficiency is associated with low birth weight infants and results in central nervous system problems (Figure 10.5).

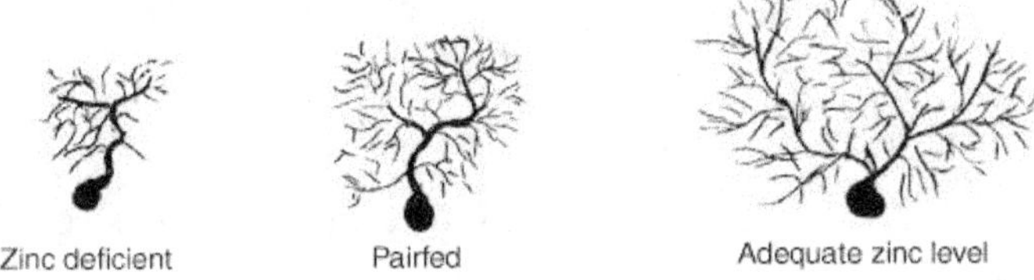

Figure 10.5   Development of Purkinje cells

7. **Skin health**   Zinc plays a role in the incorporation of the amino acid methionine into the protein of the skin.

8. **Taste**   Low blood levels of zinc have been associated with hypogeusia, in which there is a loss of a sense of the taste. Hypogeusia is usually accompanied by hyposmia, or a loss of the sense of smell.

9. **Growth**   Growth in humans suffering from zinc deficiency is severely stunted, reflecting its role in protein synthesis. A diet high in animal protein, supplemented with zinc, results in a very rapid growth response.

10. It aids in the action of insulin and hence affects glucose tolerance.

11. Zinc also helps mobilize vitamin A from its storage site in the liver and facilitates the synthesis of DNA and RNA necessary for cell reproduction.

Zinc content of some common foods is given in Table 10.3.

DEFICIENCY

1. There is retardation of growth and genital development.

2. Loss of taste acuity (hypogeusia) and hyposmia (impaired smell) are also seen.

3. Acrodermatitis enteropathica is a rare congenital defect of zinc absorption. Anorexia, diarrhoea, stunting, wasting, skin desquamation with ulceration, a reduction in lymphoid tissue and an increased susceptibility to infections are seen in this condition.

4. Skeletal development is affected.

5. Wound healing is delayed.

6. Impaired protein and nucleic acid metabolism is seen.

7. Brain development is affected.

8. Enzyme activity is affected.

Table 10.3   Food sources and their zinc content

Foodstuff	Zinc content mg/100 g
Cereals	
Rice (Raw)	1.5
Rice (Parboiled)	1.9
Wheat	3.5
Pulses	3–5
Peas	3.2
Peanuts	2.2
Oranges	0.2
Egg	1.8
Milk	0.8
Cabbage	0.8
Carrots	1.2

TOXICITY

Although intake of 2 g or more of zinc sulphate causes acute gastrointestinal problems and vomiting, there have been few reports of toxicity from its ingestion even up to 200 mg. It is suggested that overuse of zinc may cause atherogenesis (hardening of the arteries).

REQUIREMENTS

Children 1–10 year	10 mg
Above 10 years	15 mg
Pregnant women	20 mg
Lactating women	25 mg

CHROMIUM

Chromium is present in all organic matter and appears to be an essential nutrient. Only the trivalent form is biologically active.

FUNCTIONS

1. **Carbohydrate metabolism** Chromium has been identified as part of the glucose tolerance factor which is required for optimal utilization of glucose. Chromium aids in binding insulin to the cell and so in turn allows glucose to be taken up by the cell. It aids in binding insulin to the cell by forming a bridge between the insulin molecule and the membrane.

   Low intake of chromium has been associated with a reduced tolerance to glucose and an increasing incidence of diabetes both of which occur with increasing age. Many cases of mild intolerance can be treated with chromium.

2. It helps in growth.

3. **Normal amino acid metabolism** With decrease in chromium, incorporation of certain amino acids in liver and heart muscle was reduced.

4. It is necessary for maintaining normal glycogen reserves.

5. **Maintenance of normal blood cholesterol levels** Due to elevated cholesterol levels there is occurrence of increased aortic lesions and people can succumb to coronary heart diseases. Recent studies have demonstrated a significant reduction of total serum cholesterol in persons treated with chromium supplementation. There was a lowering of LDL and increase in HDL cholesterol.

METABOLISM

Cr^{6+} is absorbed more readily than Cr^{3+}. Chromium absorption is increased by oxalate. Chromium absorption is higher when carbohydrate is ingested along with chromium supplements.

Transferrin and albumin are capable of binding absorbed chromium and transporting it as part of blood serum or plasma.Other plasma proteins including α and β-globulins and lipoproteins bind chromium and thus may have a role in chromium metabolism.

Chromium is primarily excreted through the kidneys with small amounts lost in hair, sweat, and bile.

Mode of Action

Chromium potentiates insulin action and thus influences carbohydrate, lipid and protein metabolism.

- It is suggested that GTF-chromium helps to form a complex between insulin and insulin receptors that facilitates the insulin–tissue interaction.

RECOMMENDED DIETARY ALLOWANCES

Children	1–3 years	0.02–0.08 mg/day
	4–6 years	0.03–0.12 mg/day
	7 years–adults	0.05–0.20 mg/day

Insulin resistance manifested by impaired glucose tolerance has responded positively to chromium supplementation with normalized blood sugar levels.

SOURCES

Chromium is found in foods of plant sources

Vegetables	30–55 ppm
Whole grains and cereals	30–70 ppm
Fruits	20 ppm

MAGNESIUM

The presence of magnesium in living organisms was discovered in 1859. Even before that time it had long been used as a healing substance, an anaesthetic, and an anticonvulsant. By 1926 magnesium had been identified as a dietary essential for mice and by 1932 as an essential for rats. Magnesium is now recognized as an essential element that occurs predominantly within the cell. It is involved in over 300 different enzyme reactions.

The magnesium content of an infants' body at birth is approximately 0.5 g, transferred to the foetus in the latter part of pregnancy. In the adult, the magnesium content of the body is a little less than 30 g, 60% of which is concentrated in the bone. About one-third of this magnesium is closely bound with phosphate, the remaining adheres to the bone surface, from which it is readily mobilized to maintain normal blood and tissue levels.

In the blood, which contains 1% of the body magnesium, about half of the magnesium is free, one-third is bound to albumin, and the rest is part of a variety of other compounds. Magnesium occurs primarily in the RBC than in the serum (Figure 10.6).

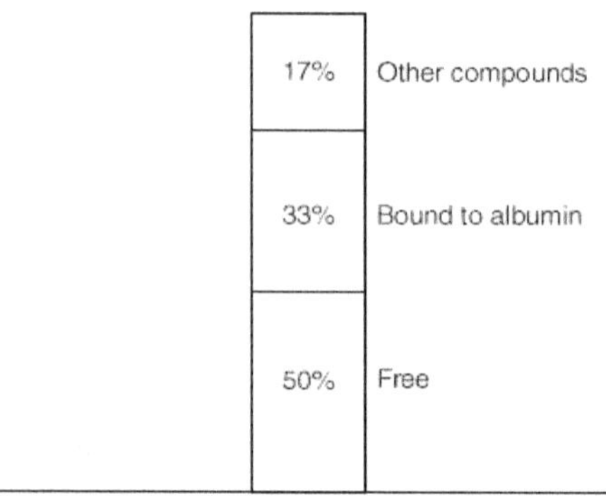

Figure 10.6   Blood magnesium

The distribution of magnesium in various compartments is summarized in the Table 10.4.

Table 10.4   Distribution and concentration of magnesium

Distribution	Percentage	Concentration
Bone	60–65%	0.5% of bone
Muscle	27%	3.5–5 mmol/kg wet weight
Other cells	6–7%	3.5–5 mmol/kg wet weight
Extracellular	<1%	1.62–2.73 mmol/L
Erythrocytes	–	1.65–2.73 mmoL/L
Serum	50–55% (as free Mg^{2+}) 13–17% as (citrate, phosphate) 32–33% (bound primarily with albumin)	0.65–0.88 mmol/L
Cerebrospinal fluid	55% (free Mg^{2+}) 45% (complexed)	1.25 mmol/L
Sweat	–	0.3 mmol/L
Secretions (Saliva, gastric bile, etc.)	–	0.3–0.7 mmol/L

FUNCTIONS

Magnesium plays a role in the biochemistry and physiology in the following way.

1. It acts as a catalyst in many biological reactions, many of which take place in the mitochondria.

2. It is involved in reactions where there is expenditure or release of energy. Some of these reactions include.

   - synthesis of fatty acids
   - activation of amino acids
   - protein synthesis
   - phosphorylation of glucose
   - oxidative decarboxylation of citrate
   - transketolase reactions

3. It is necessary for the formation of cyclic adenosine monophosphate (cAMP) which is a second messenger. It receives messages from outside the cells in the form of hormonal or other stimuli.

4. Autonomic control of the heart depends on Mg^{2+} ions in various ways:

   - Binding of neurotransmitters to their receptors
   - Coupling of receptors to adenylate cyclase
   - Activation of proteins by Mg^{2+}-dependent phosphotransferases
   - Activation of voltage-gated Ca^{2+} channels
   - Mg^{2+} regulates ion movements, e.g. the mechanism in cardiac muscle that allows K^+ to move readily into the cell but not out, is related to its blocking by intracellular Mg^{2+}.

5. Magnesium is necessary for the conduction of nerve impulses and for normal muscular contraction. In these reactions, magnesium and calcium play antagonistic roles, calcium stimulates and magnesium relaxes. The relaxing effect of magnesium is evident from the fact that with increasing levels of the element in the blood, there is an increasing anaesthetic effect. At extremely high serum magnesium levels, coma and eventually heart failure result. These levels may be reached when there is kidney failure in which excretion of magnesium is depressed.

6. Magnesium is necessary for the release of parathyroid hormone and for its action in the bone, kidney, and intestine.

7. It is also needed for the reactions involved in converting vitamin D to its active form.

ABSORPTION AND METABOLISM

Magnesium is absorbed primarily in the small intestine, probably with the help of a specific carrier. Approximately 35 to 40% of an average magnesium intake is absorbed.

Absorption is reduced in the presence of calcium, alcohol, phosphate, phytates and fat and is increased by dietary vitamin D and lactose.

The absorption of magnesium is enhanced when parathyroid hormone is secreted.

Magnesium excretion is regulated through the kidneys. When magnesium intake is low, the kidney reabsorbs almost all magnesium so that practically none is lost. Urinary losses of magnesium increases with the use of diuretics and consumption of alcohol. Dietary unabsorbed magnesium is excreted in the faeces. The amount lost in perspiration is usually small, but at high temperatures it may amount to 15% of magnesium losses. The metabolism of magnesium is controlled by the thyroid gland.

DEFICIENCY

Deficiency occurs when there is starvation, persistent vomiting, trauma of surgery, very high calcium intake or rapid transit of food through the gastrointestinal tract, which reduces absorption time.

Symptoms of deficiency are, irritability, nervousness, convulsions as a result of stimulation of nerve impulses and increased muscular contraction.

There is low magnesium tetany, initially is seen as uncontrolled neuromuscular tremors that progress until convulsive seizures occur.

Calcification of soft tissues occur since there is increased calcium absorption in magnesium deficiency.

Inadequate magnesium also results in vasodilation and skin changes.

REQUIREMENT

Adult men and women	300–350 mg/day
Children	100–250 mg/day
Adolescents	200–300 mg/day

FOOD SOURCES

Vegetables, legumes, seafood, nuts, cereals and dairy products are good sources. Green leafy vegetables are also good sources.

ASSAY

The amount of magnesium excreted in the urine after a large oral dose is the most effective and sensitive method of assay. Retention of more than 40% of dietary magnesium indicates that the body needs to correct a deficiency.

SELENIUM

The highest concentrations of selenium in the body are found in liver, kidney, heart and spleen. It is deposited in all body tissues except fat. Serum levels are about 0.22 μg/dl.

FUNCTIONS

1. There is an overlap between function of selenium and vitamin E.

   Selenium is an integral part of the antioxidant enzyme glutathione peroxidase which inactivates the enzymes that cause oxidation or rancidity in fats or which cause oxidative damage to cell membranes. Glutathione peroxidase is a seleno protein which catalyses peroxidation of glutathione. This enzyme is a protective agent against accumulation of organic peroxidases and hydrogen peroxide within the cell. Selenium is part of glutathione peroxidase which changes the lipid peroxides that are formed to alcohol and water.

   Vitamin E is also an antioxidant and has a similar function.

2. Selenium is incorporated into the protein of the teeth.

3. It is required for proper liver functioning.

4. It helps in the release of energy to cells.

5. It plays a part in the development of structural protein of sperm cell and is therefore essential for normal growth and fertility.

6. It reduces incidences of cancer and heart disease because of its antioxidant function.

7. It prevents growth retardation.

8. It protects red cell membrane and haemoglobin against oxidative changes.

9. It protects body against toxicity due to mercury and cadmium.

SELENIUM DEFICIENCY

Keshan's disease is a congestive heart disease in which heart muscles of children and women of child bearing age undergo degenerative changes and can be cured by selenium supplements.

Patients on IV feeding devoid of selenium have shown muscular discomfort, finger-nail defects and biochemical abnormalities in RBC.

REQUIREMENT

250–300 μg of selenium provides protection against cancer and heart disease but this is not an established fact. Because of its role in glutathione peroxidase, selenium probably interacts with any nutrient that affects the antioxidant/pro-oxidant balance of the cell. For example selenium requirement of chicks is inversely proportional to the dietary vitamin E intake. Selenium also protects against the toxicity of mercury, cadmium and silver.

ABSORPTION, TRANSPORT AND EXCRETION

It is completely absorbed as selenomethionine and generally well absorbed as inorganic forms and ranges from 50–100% of intake.

Transport

It is found in plasma as selenoprotein P and a glycosylated form of glutathione peroxidase which contains selenium as selenocysteine.

Excretion

Homoeostasis of selenium is achieved through regulation of excretion and is usually excreted through the urine. Metabolism of selenium is summarized in Figure 10.7.

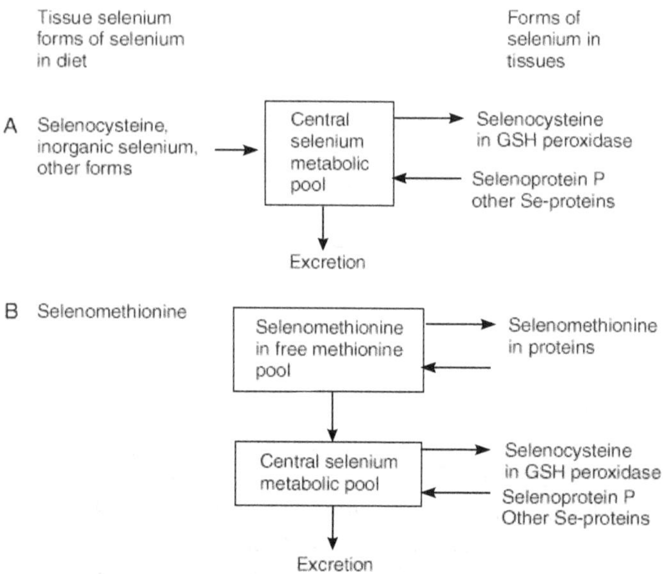

Figure 10.7   Metabolism of selenium

○ REVIEW QUESTIONS

1. List the functions of iron.

2. What are the factors affecting iron absorption?

3. What are the symptoms of iron deficiency?

4. Give the RDA for iron and iodine.

5. Give the symptoms of goitre.

6. What are goitrogens?

7. Explain the functions of iodine.

8. Explain the grades of IDD.

9. Explain the functions and deficiency of copper.

10. Explain the sources, deficiency and excess of fluorine in humans.

11. Give the relationship between glucose and chromium.

12. Explain selenium–vitamin E relationship.

13. Explain the functions and deficiency of zinc.

CRITICAL THINKING QUESTIONS

1. What is the effect of consuming too much iron?

2. How reliable is hair analysis for the assessment of zinc?

11

ANTIOXIDANTS

- Free radical formation increases during diabetes, exercise, diseases, etc.

- In biological systems, removal or addition of electrons is known as redox function.

- Free radicals and reactive oxygen species abstract hydrogen from membrane phospholipids which contain polyunsaturated fatty acid chains, producing free radicals.

- Oxidatively modified protein of the lipoprotein LDL is known to be the mediator in cholesterol deposition in the blood vessels.

- Vitamin E, C, carotenoids, oestrogen, flavonoids, etc. are known as primary antioxidants.

- Copper, gluthathione reductase, transferrin, ceruloplasmin, etc. are known as secondary antioxidants.

- Vitamin E is an example of a phenolic antioxidant.

- Dietary antioxidants include β-carotene, vitamin C, riboflavin, sulphur containing amino acids, selenium, etc.

Oxidative stress has been defined as a disturbance in the equilibrium status of pro-oxidant/antioxidant systems in intact cells. This means that cells have intact pro-oxidant/antioxidant systems that function continuously to generate and to detoxify oxidant during normal aerobic metabolism. When additional oxidative events occur, the pro-oxidant systems may outbalance the antioxidant, resulting in oxidative damage to lipids, proteins, carbohydrates and nucleic acids. Ultimately, cell death may occur from severe oxidative stress. Oxidative stress may also induce a rapid alteration in the antioxidant systems by inducing proteins that participate in these systems and by depleting cellular stores of antioxidant materials such as glutathione and vitamin E.

A disturbance in pro-oxidant/antioxidant systems results from a myriad of different oxidative challenges, including radiation, xenobiotic metabolism of environmental pollutants and other materials, and challenges to the immune system in human disease or in abnormal immune function. A variety of radical species in these processes has led to considerable interest in the reactions of partially reduced oxygen species and radical and non-radical species derived from them. The radical may be a very small molecule such as oxygen or it may be a part of a large biomolecule such as a protein, carbohydrate, lipid or nucleic acid. Some radical species are very reactive while others are relatively inert.

RADICALS AND NON-RADICALS IN OXIDATIVE STRESS

Radicals of oxygen (superoxide anion, and hydroxyl, alkyoxyl and peroxyl radicals) reactive non-radical oxygen species (hydrogen peroxide and singlet oxygen) and radicals of carbon, nitrogen and sulphur constitute the variety of reactive molecules that cause an oxidative stress to cells. It has been estimated that a maximum of 5% of the total oxygen metabolism of liver cells results in the production of partially reduced oxygen species. This is a significant stress by itself, but extracellular sources of these molecules may be even more significant.

Some of the molecular species of oxygen in oxidative stress include the following.

Triplet oxygen (O_3)

Superoxide anion ($O_2^{-\bullet}$)

Hydrogen peroxide (H_2O_2)

Hydroxyl radical ($HO^\bullet$)

Water (H_2O)

Singlet oxygen (O_2)

Superoxide anion is generated continuously by several cellular processes, the most important of which are the microsomal and mitochondrial electron transport systems. In addition, many cellular oxidases may be important sources of this molecule. Myeloid cells have a special role in production of superoxide anion because they contain a plasma membrane-bound electron transfer complex that reduces oxygen with the reduced form of nicotinamide adenine dinucleotide phosphate (NADP) to produce copious amounts of superoxide anion.

The presence of superoxide dismutase in both cytoplasm and mitochondria ensures that much of the superoxide is rapidly converted into hydrogen peroxide. Superoxide anion which is not a particularly reactive molecule, can diffuse considerable distances from its site of production. It must be transported across membranes (by an anion transport mechanism) and in the vicinity of membranes, it may be protonated to HO_2, becoming a much more reactive substance.

Hydrogen peroxide is generated by the same sources that produce superoxide anion because both enzymatic (superoxide dismutase) and non-enzymatic destruction of superoxide anion produces hydrogen peroxide. A number of other specific enzymes also produce hydrogen peroxide directly. These include peroxisomal enzymes associated with fatty acid metabolism and cytoplasmic enzymes responsible for the oxidation of a variety of cell metabolites. Hydrogen peroxide can diffuse over considerable distances and may pass through membranes readily. Thus intracellular pools of hydrogen peroxide equilibrate rapidly across membrane boundaries.

Hydrogen peroxides and superoxide anion are found in extracellular space and in blood plasma as a result of the membrane-associated reaction of myeloid cells such as neutrophils and macrophages. The membrane-associated NADPH-oxidase produces superoxide anion that rapidly dismutes to hydrogen peroxide as well.

EFFECTS OF OXIDANTS ON MACROMOLECULES

Carbohydrates

Hydroxyl radicals react with carbohydrates by randomly abstracting a hydrogen atom from one of the carbon atoms, producing a carbon-centred radical. This leads to chain breaks in important molecules such as hyaluronic acid (Figure 11.1). In the synovial fluid of the joints, an accumulation and activation of acids during inflammation produces significant amount of oxy radicals. This phenomenon apparently accounts for a significant decrease in the synovial fluid of affected joints.

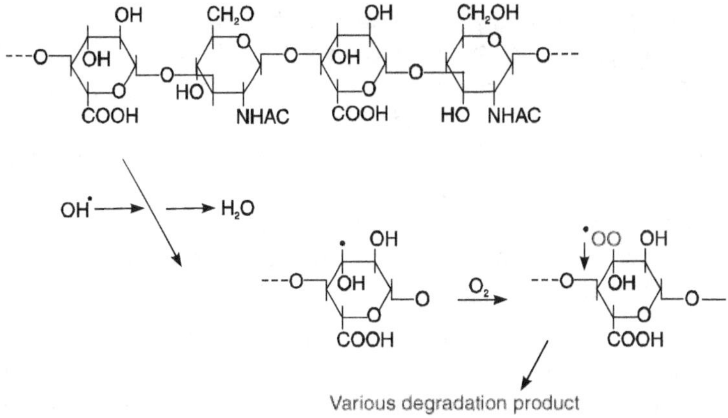

Figure 11.1   Reaction of hydroxyl radicals with hyaluronic acid

Nucleic Acids

Nucleic acids are also carbohydrate polymers that can undergo reactions with hydroxyl radicals such as those depicted for hyaluronic acid. Base modification occurs during oxidative stress. These base modifications may be responsible for genetic defects produced by oxidative stress.

An important metabolic effect of DNA damage is the rapid induction of polyadenosine diphosphate ribose synthesis (ADP ribosylation) in nuclei, resulting in extensive depletion of cellular NADH pools. ADP ribosylation has been associated with repair of damaged DNA (Figure 11.2).

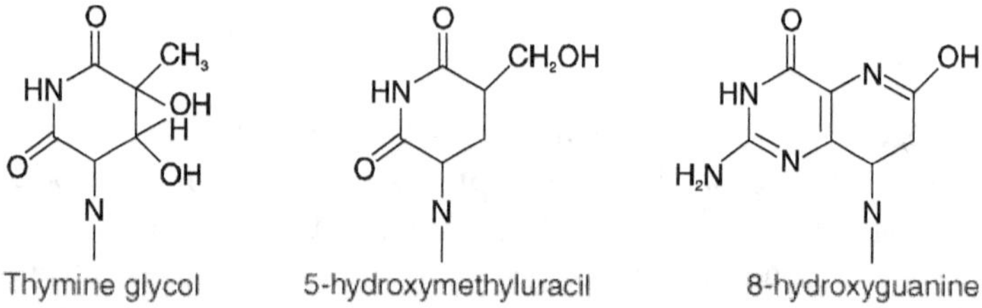

Thymine glycol   5-hydroxymethyluracil   8-hydroxyguanine

Figure 11.2 Some modified bases found in DNA after oxidative stress

Proteins

Proteins have many reactive sites that can be damaged during oxidative stress. Three events can be seen. First, aggressive radicals such as hydroxyl radical can fragment proteins in plasma. This fragmentation is associated with reactions at specific amino acids such as proline and histidine. Second, proteins may contain metal-binding sites that are especially susceptible to oxidative events through interaction with the metals. These reactions produce irreversible modifications in amino acids that might be involved in metal ion binding, such as histidine.

Finally, many intracellular proteins have "reactive" sulphydryl groups on specific cysteine residues that can be modified (oxidized) to specific forms (disulphides) which can be reduced again by metabolic processes.

Lipids

Lipid peroxidation is a facile process with polyunsaturated lipids, materials that are prevalent in dietary constituents and in cellular membranes (Figure 11.3). Peroxidation of cellular lipid results in a variety of deleterious effects on important membrane functions. Most peroxidized lipid occurs as a result of oxidative stress in intact cells; but some peroxidized lipid in the diet may be directly incorporated into cell structures. Lipid peroxidation is a radical-initiated chain reaction that is self-propagating in cellular membranes. Hence, isolated oxidative events may have profound effects on membrane function.

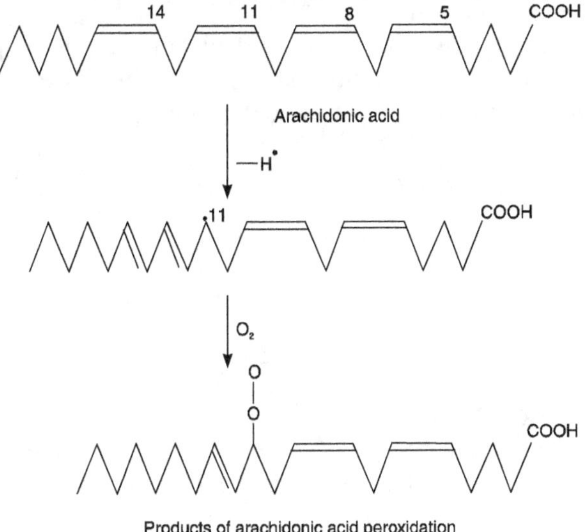

Figure 11.3   Products of arachidonic acid peroxidation

REACTIONS OF LIPID PEROXIDATION

Initiation

$$\text{LIPID} + \begin{bmatrix} R^\bullet \\ OH^\bullet \end{bmatrix} \rightarrow \text{LIPID}^\bullet + \begin{bmatrix} RH \\ H_2O \end{bmatrix}$$

Propagation

$$LIPID^{\bullet} + O_2 \rightarrow LIPID\text{-}OO^{\bullet}$$

$$LIPID\text{-}OO^{\bullet} + LIPID \rightarrow LIPID\text{-}OOH + LIPID^{\bullet}$$

Termination

$$LIPID^{\bullet} + LIPID^{\bullet} \rightarrow LIPID\text{-}LIPID$$

$$LIPID\text{-}OO^{\bullet} + LIPID\text{-}OO^{\bullet} \rightarrow LIPID\text{-}OO\text{-}LIPID + O_2$$

$$LIPID\text{-}OO^{\bullet} + LIPID^{\bullet} \rightarrow LIPID\text{-}OO\text{-}LIPID$$

Scavenging

$$LIPID^{\bullet} + VITE \rightarrow LIPID + VITE^{\bullet}$$

Malondialdehyde can be measured in the blood plasma and is used to evaluate oxidative stress. Vitamin E is very effective as an antioxidant in lipid peroxidizing systems.

Malondialdehyde (MDA)

CELLULAR ANTIOXIDANTS

The most effective antioxidant in oxidative stress depends on the specific molecules causing the stress (e.g. superoxide anion, lipid peroxides, iron-generated hydroxy radical) and the extracellular or cellular location of the source of these molecules. As an example, damage to a cell membrane occurs from both internally and externally generated oxidative stress. This damage is most effectively prevented by vitamin E which reacts with peroxyl and hydroxyl radicals; by carotenoids, which react with singlet oxygen; and possibly by membrane-bound proteins. The chain-breaking antioxidant function of vitamin E in membranes results from its close association with polyunsaturated components of the membrane. Vitamin E radical can be reduced by cytoplasmic vitamin C and glutathione or by membrane-bound quinols. Vitamin C is reduced by glutathione (Figure 11.4). Thus, a specific attack on membranes results in the participation of at least three different antioxidants. Similarly, when oxidative stress occurs in plasma, a variety of different antioxidants participate in the response.

Glutathione and Protein Sulphydryls

The low molecular weight thiol, glutathione and "reactive" protein sulphydryls (specific cysteines in many proteins) are primary participants in cellular antioxidant systems. The enzymes of the glutathione redox cycle are shown in the Figure 11.4.

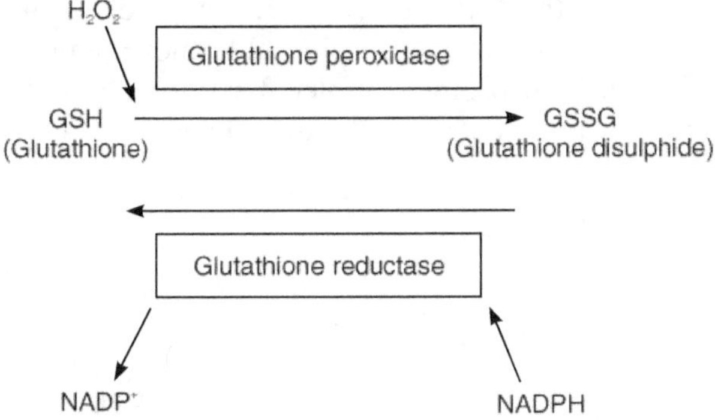

Figure 11.4   Enzyme of glutathione redox cycle

In this cycle, glutathione is oxidized by hydrogen peroxide to glutathione disulphide by the selenium-containing enzyme, glutathione peroxidase, and by other enzymes. Thus glutathione can detoxify both soluble and lipid peroxides. Glutathione disulphide is subsequently reduced by glutathione reductase, using NADPH as the reductant produced by the pentose phosphate pathway. Glutathione also acts as a reductant for vitamin C and it can directly reduce some protein-bound sulphydryls. In combination with an enzyme called glutaredoxin (or thiol transferase), glutathione can reduce a larger number of protein sulphydryls.

The concentration of cellular glutathione has a major effect on its antioxidant function. Nutrient limitation, exercise, and oxidative stress have major effects on glutathione concentration.

Vitamin E and Membrane Peroxidation

Vitamin E has hydroxylated aromatic rings (chromanol rings) and polyisoprenoid side chains. The molecule is highly lipophilic and resides almost exclusively in cell membranes. The chromanol ring may be at the surface of the membrane and the polyisoprenoid chain is inserted into the bilayer. The chromanol ring is the active radical-quenching part of the vitamin. Because lipid peroxidation occurs on unsaturated fatty acid chains that reside within the lipid bilayer, the action of vitamin E as an antioxidant must involve considerable movement of the lipids and vitamin E to promote molecular interaction. The reaction that takes place is shown in Figure 11.5.

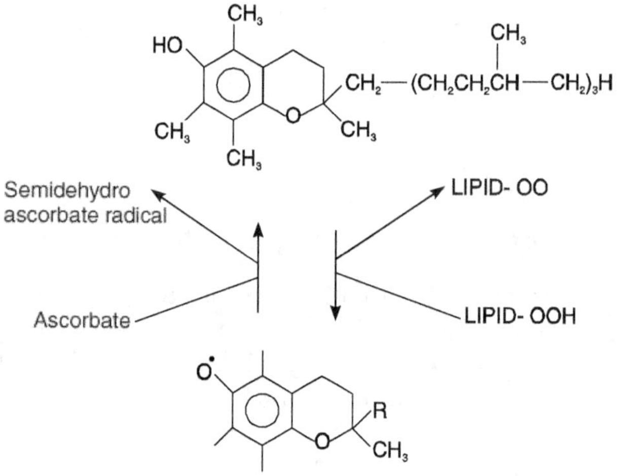

Figure 11.5   Reaction of vitamin E with lipid radicals

Vitamin E is a chain-reaction-breaking antioxidant because it quenches the intermediate in the chain reaction. The ascorbate radical formed in this process reacts rapidly with reduced glutathione.

Enzymes

Superoxide dismutase is one of the most important enzymes that function as cellular antioxidant. It is present in cell cytoplasm (copper–zinc enzyme) and in mitochondria (manganese enzyme) in order to maintain a low concentration of superoxide anion.

$$2O_2^- + 2H^+ \xrightarrow{\text{Superoxide dismutase}} O_2 + H_2O_2$$

The absence of this enzyme is lethal, but an increase in its concentration may not increase antioxidant protection.

Catalase is a haem protein, usually found in peroxisomes, which plays a protective role that is similar to that of glutathione peroxidase in most cells.

$$2H_2O_2 \xrightarrow{\text{Catalase}} O_2 + 2H_2O \text{ catalase}$$

Superoxide dismutase and catalase provide a rapid means of equilibrating and detoxifying superoxide anion and hydrogen peroxide in cells.

Plasma Antioxidants

The antioxidant properties of this important fluid resides primarily in a number of small molecules and protein constituents. Ascorbate is among the first compounds that become oxidized in stress. Other plasma components may become oxidized only when ascorbate is depleted. Bilirubin (bound to albumin) and uric acid, both considered to be metabolic waste products in plasma, are potentially good scavengers of oxyradicals.

The antioxidant proteins of plasma that are most important include ceruloplasmin, albumin, transferrin, haptoglobin and haemopexin.

Oxidative stress also affects plasma lipid particles such as low-density lipoproteins (LDL). The protein and lipids in LDL are good targets for oxidation and the oxidized forms of LDL are strongly implicated in the formation of fatty lesions (atheromas) in artery walls. The apoprotein B component of these particles is fragmented by oxidation. LDL particles contain a significant amount of vitamin E and carotenoids that may serve as primary antioxidants.

HUMAN DISEASE AND OXIDATIVE STRESS

Cancer

Radicals of differing kinds, including oxyradicals are involved in both initiation and promotion in multistage cancer development. In this process DNA is damaged and the cellular antioxidant systems are modified as a result of the expression of different genetic components in precancerous and tumour cells. Because specific genes are apparently controlled by oxidation/reduction switching of important gene regulatory proteins, the effect of oxidative stress may be manifested directly by alterations in these specific proteins.

Some of the anticancer drugs are also promoters of oxyradical production. These drugs may be effective because of their ability to generate oxidative species that cause damage of DNA, membrane or enzyme in tumour cells.

Cataract

The crystalline, the major proteins of the eye lens are long-lived proteins that are unusually abundant in methionine and cysteine groups. These are easily oxidized components of the protein during oxidative stress. The lens has a high concentration of both glutathione and glutathione reductase. Glutathione reduces substantially in lesions of the lens. The vitreous humor of the eye also contains hyaluronic acid that is depolymerized when exposed to oxyradicals.

Arthritis and Rheumatic Disorders

These diseases are characterized by inflammatory responses in which extensive tissue damage can occur through oxidative stress due to oxidation of hyaluronic acid.

NUTRITIONAL EFFECTS ON OXIDATIVE STRESS

Minerals such as selenium, iron, copper and zinc, vitamins such as A, C and E and other supplementary materials such as carotenoids, cholesterol and unsaturated fats play important roles in the balance between pro-oxidant and antioxidant systems in humans.

Alcohol toxicity results in a direct increase in cellular oxidative stress. Ethanol produces lipid peroxidation in liver.

Dietary polyunsaturated fatty acids may also have a direct effect on the peroxidation of cellular lipids because these dietary lipids can alter the membrane composition of various cells, making them more or less susceptible to peroxidation damage during oxidative stress.

REVIEW QUESTIONS

1. Define antioxidant.

2. Explain the role of antioxidant in combating free radicals.

3. How do antioxidants prevent disease?

4. What is oxidative stress?

5. Discuss the effect of oxidants on macromolecules.

6. What are some of the modified bases found in DNA after oxidative stress?

7. Write a note on the reactions of lipid peroxidation.

8. What are cellular antioxidants?

CRITICAL THINKING QUESTION

1. Is vitamin E the only antioxidant?

12

FLUID ELECTROLYTE HOMEOSTASIS

Did You Know?

- Vitamins often need minerals to help perform metabolic reactions. For example, the coenzyme for the vitamin thiamine (TPP) requires magnesium to function efficiently.

- Humans can survive longer without food than without water. The body can conserve water, but some must be lost every day. Thus after only a few days, severe dehydration and death can result from lack of water intake. Death from starvation can take as long as 50 or more days in an adult.

- Water evaporation from the skin requires heat energy. So when perspiration evaporates, heat energy is taken from the skin, leaving you feeling cooler. Perspiration on the skin tastes salty, not because it has a higher sodium concentration than blood, but because once the water evaporates, concentrated sodium is left behind.

- The estimation of dietary water needs is 1 millilitre per kilocalorie.

- Sodium is often added to proceed foods. Sodium is prevalent in foods such as frozen dinners, canned soups, and convenience entrees.

- Oranges are alkaline ash foods. Most fruits and vegetables are metabolized, the potassium and sodium left behind form alkaline compounds. The body's mechanisms that regulate acid–base balance, however, can compensate for the alkaline products formed.

- A preference for salty foods is partially learned. Although part of our preference for salt is related to the sodium receptors on the tongue, habit also has a role. If you grow up consuming salty foods, you are more likely to prefer salty foods. Preferences can be relearned by gradually decreasing added salt.

WATER

All the fluids in the body have water as the medium. Water is of fundamental importance to all tissues, both structurally and functionally. Deprivation of water causes death much more quickly than deprivation of food. Most tissues contain more than 70% of water; even bone is nearly one-third water. Adipose tissue having the least amount, possesses appreciable quantities in connective tissue and in spaces between fat cells. About half of body water is found in muscles. In a man of average build, water makes up nearly 60–70% of the body weight, intracellular fluid making about 40–50% and extracellular fluid about 20%; of the latter, interstitial fluids provide nearly 17% and intravascular fluid (blood plasma) 3–5%.

In a healthy adult, the body contains on an average 42–45 litres of water of which 25–30 litres are present in intracellular fluids. Extracellular fluids contain 14–16 litres, of which blood plasma contributes 3–3.5 litres, lymph about 1–1.5 litres and interstitial fluids, 11–12 litres.

Water has some unique properties. It has high values for boiling point, melting point, dielectric constant and specific heat. Its density is highest at 4°C. Because of high heat capacity of water, the body can undergo large changes in heat production with very little alteration in body temperature. A constant circulation of water in blood is an important factor in maintaining the constant temperature of the body. As the latent heat of evaporation for water is high, the loss of small amounts of water by evaporation of sweat means a relatively large heat loss. The high latent heat of solidification protects against freezing of the tissues. Water is a good solvent, and plays a fundamental role in cellular reactions. Various substances in tissues are in solution in water both intracellular and extracellular. There are true solution of electrolytes, like NaCl, or non-electrolytes like glucose as also colloidal solutions of large molecules, like proteins. Fats and fat-soluble substances which are insoluble in water can be carried in it as fine emulsions, or can be rendered water-soluble by combination with lyophilic substances.

There is an approximate state of balance between the input of water into the body and its output from the body, the total quantity inside the body normally remaining fairly constant.

Table 12.1   Water balance in healthy adult

Input/day (ml)		Output/day (ml)		Water pool in the body (ml)	
As drinks	1500	Through urine	1500	Plasma	3000
With food	700	Through skin	400	Tissue	12000
From metabolism (oxidation)	300	Through lungs	400	Cell	30000
		Through faeces	200		
	2500		2500		45000

As shown in Table 12.1, a drink includes water, tea, milk or any other aqueous fluids taken directly. Food water represents that in cooked food. Metabolic water is provided by the chemical processes mainly oxidation occurring inside the body. Urine output is variable, depending on intake. Perspiration may be visible, but there is also enough invisible evaporation from the skin. Expired water depends on metabolic function and activity and faeces water on digestive and absorptive processes.

The proportion of water in the intracellular and extracellular compartments of the body is maintained by osmotic pressure of plasma (usually termed colloid osmotic pressure as it is mainly due to the proteins in the colloidal state) on one side and the capillary arterial pressure on the other. The process is a constant exchange of water between blood and tissues via the lymph.

ELECTROLYTES IN THE BODY

There are some electrolytes in solution in body water. The average normal values of cations and anions in body fluids is presented in Table 12.2. The values presented show that each compartment maintains its electrical neutrality by balancing the cations and anions. Values in plasma do not differ significantly from those in all extracellular fluids taken together. But the intracellular fluid has widely different values in the proportion as well as in the total.

However, almost the whole of proteins of extracellular fluids is found in plasma. Cerebrospinal fluid compares with plasma with slightly higher value for Na^+ and lower value for K^+/Ca^{2+} is half that in plasma; Cl^- is higher but HCO_3^- is the same.

Table 12.2　Average normal concentration of cations and anions in body fluids (in mEq/l)

	Plasma	Extracellular fluid	Intracellular fluid	Cerebrospinal fluid
Cations				
Na^+	140	145	10	150
K^+	5	5	150	3.5
Ca^{2+}	5	2	2	2.5
Mg^{2+}	3	2	15	–
Anions				
Cl^-	100	100	10	125
HCO_3^-	28	28	10	26
PO_4^{3-}	2	2	90	–
SO_4^{2-}	1	1	15	–
Proteins	16	18	52	300 mg/l
Others	6	5	–	–

HYDROGEN ION CONCENTRATION AND BUFFERS

The concentration of hydrogen ion in plasma is only 10^{-7} g ions/l which is insignificant compared to other cations, and thus has little or no influence in maintaining the electrolytic balance. But its changes have profound effect on health. The hydrogen ion concentration is represented by pH and blood maintains a constant pH even though there are some changes in the concentration of acidic or basic substances.

Phosphate Buffer System

At the blood pH, H_2PO_4 can act only as an acid and HPO_4^{2-} only as a base because their functioning in the reverse process would require a pH far removed from it.

Protein Buffer System

Because of the high concentration, proteins are the most important buffers in body fluids. The dissociation of the imidazolium group of the histidine residue of a protein molecule can occur efficiently at the blood pH and proteins act as good buffers. In a protein molecule, there may be a number of histidine residues. The imidazolium group of each has different pH optima for dissociation which corresponds to regions in health and disease.

In the red cells, haemoglobin acts as the most important buffer by virtue of its high concentration and presence of 36 histidine residues in its globin component. Moreover both oxygenated haemoglobin ($H.HbO_2$) and non-oxygenated haemoglobin (H.Hb) can dissociate as below:

$$H.HbO_2 \;\square\; H^+ + HbO_2^-$$

$$H.Hb \;\square\; H^+ + Hb^-$$

The buffering capacity of haemoglobin is attributed to the effect of the imidazole groups which are constituents of the histidine component of the globin.

Thus if the haemoglobin is more oxygenated the imidazole groups become more acidic and therefore more dissociated and hence the buffering capacity is also increased. Oxygenated haemoglobin is (HbO_2) a stronger acid than reduced haemoglobin.

Proteins in the plasma are also involved in a buffering action but to a minor extent.

$$Protein + H^+ \rightleftharpoons H^+ Protein$$

$$HPO_4^- + H^+ \rightleftharpoons H_2PO_4^-$$

If a fixed base, e.g. NaOH, enters the extracellular fluid it reacts with the acid component of these buffer systems, chiefly H_2CO_3 as follows.

$$H_2CO_3 + NaOH \rightleftharpoons NaHCO_3 + H_2O$$

$$H^+ protein + NaOH \rightleftharpoons Na\ protein + H_2O$$

$$H_2PO_4^- + NaOH \rightleftharpoons NaHPO_4^- + H_2O$$

CO_2 is buffered only to a lesser extent by the plasma protein buffer system.

CO_2 forms H_2CO_3

$$CO_2 + H_2O \rightleftharpoons H_2CO_3$$

$$H_2CO_3 + B\ protein \rightleftharpoons BHCO_3 + H^+ protein$$

H^+ protein is a weaker acid compared to H_2CO_3.

Factors Influencing Distribution of Body Fluid

The osmotic forces maintained by the solutes is one of the major factors which helps control the distribution of body fluid.

1. Organic compounds having a large molecular size, e.g. proteins. They help in the exchange of fluid between the circulating blood and the interstitial fluid.

2. The inorganic electrolytes like various cations and anions of the body fluids are important factors in directing the movement of the fluid in the various compartments. The main cations are Na^+ and K^+ ions. Ca^{++} and Mg^{++} are

present in smaller amounts. The main anions are Cl^- and HCO_3^- the inorganic electrolytes.

FUNCTIONS

The functions of water are as follows:

Water is a part of all tissues and is essential for growth. Fat tissue is one-fifth water and muscle is three-fourths water. The cell water and its content in solution provide a normal turgor or fullness to the tissues, and a distension or degree of rigidity of the cell contents on the cell membranes.

Water functions as a solvent; many metabolic reactions of life are made possible due to this property of water.

Water also plays an important role in the distribution of heat throughout the body and in the regulation of body temperature.

SODIUM, POTASSIUM AND CHLORIDE METABOLISM

The metabolism of sodium, potassium and chloride are closely related to one another, the three elements are studied together. Sodium is the main cation of the extracellular fluid (ECF) and potassium the main cation of the intracellular fluid (ICF). Potassium is also an important constituent of the ECF though it is present in lesser concentration. Chloride forms the chief anion of the ECF and exists along with sodium.

DISTRIBUTION

The strict localization of large amounts of sodium and chloride in the ECF and potassium in the ICF is mainly due to the high degree of selective impermeability of the cell membrane (*see* Table 12.2 for composition of ECF and ICF) to these ions, under normal conditions. This pattern may deviate from the normal, under certain functional and pathological states. In such cases, potassium will move from the ICF to ECF, while sodium from ECF moves into the cells. These changes are always accompanied by alteration in the concentration of chloride and bicarbonate in the ECF.

The distribution of Cl^- and HCO_3^- between plasma and RBC is identical.

FUNCTIONS

1. Sodium, potassium and chloride regulate the acid–base balance of body fluids. Chloride is required for the chloride shift mechanism and for the formation of HCl in gastric juice.

2. They regulate the water balance by maintaining the osmotic pressure of the body fluids.

3. They help to preserve the normal neuromuscular irritability, by maintaining a state of equilibrium on account of their relative proportion in the ECF and ICF.

4. Sodium, potassium and chloride help in the preservation of the permeability of the cells.

5. The plasma proteins are kept in solution as proteinates by sodium and potassium, because both albumin and globulin are soluble in dilute neutral solution of salts (NaCl).

Source

Sodium and chloride The main source is as salt which is used in cooking. Sodium chloride is present in plant and animal tissues. The recommended amount of sodium for adults is 5 to 15 g per day.

Potassium Vegetables are rich sources of potassium. The recommended amount of potassium for adults is about 4 g/day.

ABSORPTION AND EXCRETION

Sodium, potassium and chloride are absorbed readily from the gastrointestinal tract.

Much NaCl is not lost through the skin under normal conditions. However, during prolonged strenuous exercise, excessive sweat is accompanied by considerable loss of NaCl with onset of symptoms like muscle cramps, headache and mental confusion.

Potassium is excreted mainly through the kidneys. Some potassium is also secreted by the distal tubules and this adds to the quantity of potassium in urine.

SODIUM AND CHLORIDE METABOLISM

The metabolism of sodium and chloride are so much interrelated that abnormality in the metabolism of one is always accompanied by the abnormality in the other. Thus loss of sodium in gastrointestinal fluids, sweat and urine is accompanied by loss of chloride also, though there is more loss of chloride than sodium. This produces a low level of plasma chloride and a compensatory increase in bicarbonate with consequent alkalosis.

The metabolism of sodium is influenced by the adrenocortical hormones. In adrenocortical insufficiency there is increased excretion of sodium in the urine and a reduced serum sodium concentration. This is called hyponatremia. Hyponatraemia can occur due to a loss of sodium or due to overhydration.

Hypernatraemia means increased serum sodium level which occurs during hyperactivity of adrenal cortex as in Cushing's disease.

POTASSIUM METABOLISM

Potassium is the main cation of the intracellular fluid and is also an important constituent of the ECF. Potassium is essential for growth and for building up of cells. This is indicated by a fairly high retention of potassium during infancy, childhood and pregnancy. Potassium ions are essential for contraction of cardiac and skeletal muscles.

Hyperkalaemia means increased serum potassium. Hyperkalaemia occurs in renal failure, advanced dehydration, Addision's disease, etc. Hypokalaemia means decreased serum potassium concentration. Prolonged diarrhoea, vomiting, Cushing's syndrome, etc. cause hypokalaemia.

REVIEW QUESTIONS

1. Give the importance of water in the body.

2. What is oedema?

3. Explain how water balance is maintained.

4. Explain the role of hormones in maintaining electrolyte balance.

5. What is hypokalaemia and hyponatraemia.

CRITICAL THINKING QUESTION

1. Is water balance affected by extreme heat?

13

HORMONE AND NUTRIENT INTERACTIONS

- Hormones are chemicals that are secreted into the blood by endocrine glands.

- Chemical classes of hormones include amines, polypeptides, glycoproteins and steroids.

- β-cells in the islets of Langerhans of the pancreas secrete insulin while the α-cells secrete glucagon.

- Insulin lowers blood glucose and stimulates the production of glycogen, fat and protein.

- Thyroxine and epinephrine stimulate the rate of cell respiration in the body.

- Hormones from the adipose tissue contributes to the sensation of hunger.

- Parathyroid hormone and calcitonin regulate calcium deposition and resorption in bones.

The various hormones produced by the endocrine glands has a profound control on the metabolism of many nutrients and many of the immune cells function as hormone-elaborating cells that mediate changes in metabolism during severe illness. Hormones also affect nutrient storage and mobilization.

EFFECT OF PANCREATIC HORMONES

Carbohydrate Metabolism

Carbohydrate metabolism is finely regulated by interactions between insulin, the hormone promoting fuel storage and the counter-regulatory hormones such as glucagon, epinephrine, cortisol and growth hormone. Because the brain requires a constant supply of glucose even in the absence of available carbohydrate, the homeostatic mechanisms are very important.

Insulin plays the central role in regulating glucose metabolism. Insulin initiates its metabolic effect by binding to a cell-surface receptor. These effects depend on the activation of a tyrosine-specific protein kinase which is contained in the β-subunit of the receptor.

After binding to the cell-surface receptor, insulin accelerates the membrane transport sugars. Hence glucose transport increases many times the original value. This is due to the translocation of glucose transporter proteins from intracellular membrane pools to plasma membrane in insulin-sensitive tissue. This translocation is mediated by insulin.

Transporters located in the insulin-sensitive tissues, i.e., adipose tissue, skeletal muscle and heart muscle are sodium-independent; facilitative transporters are the primary proteins involved in insulin-stimulated translocation.

Many tissues, including brain and liver, maintain glucose uptake independent of insulin concentration. The hypothalamus may represent a brain region with some dependence on insulin for glucose use, particularly in the glucose-sensitive areas of the ventromedial and lateral nuclei. Conversely, skeletal muscle (generally an "insulin-sensitive" tissue) may take up glucose without insulin stimulation under circumstances of contractile stimulation.

In addition to increasing glucose transport, insulin also has major effects on intracellular glucose metabolism. In diabetes, activities of enzymes involved in glycolysis and glucose oxidation, such as glucokinase, phosphofructokinase and pyruvate kinase are decreased and activities of gluconeogenic enzymes, such as glucose 6-phosphatase, fructose 1,6- biphosphatase, phosphoenol pyruvate carboxykinase and pyruvate carboxylase are increased. These abnormalities do not occur when adequate insulin is present. Insulin also promotes glycogen synthesis by promoting the conversion of glycogen synthase to its active, glucose 6-phosphate-independent form and by decreasing the activity of phosphorylase.

The ingestion of carbohydrate produces a prompt increase in plasma insulin and a decrease in glucagon concentration. The rise in insulin occurs before the rise in arterial glucose concentrations. This increased insulin release creates a "priming effect" in which the action of insulin begins concurrently with the absorption of glucose to minimize the extent of hyperglycaemia after a meal.

During periods of starvation, the maintenance of euglycaemia is critically important to the organism in the non-ketotic state. The energy needs of the brain can only be met by glucose, and its absence results in the death of the central nervous system tissues. The glucose pool can provide only 15 to 20 g in the adult. Hepatic glycogen can be mobilized to provide circulating glucose of 70 g. The stored glycogen can only provide for less than 8 hours of glucose on average. Thus gluconeogenesis is important for the maintenance of post absorptive plasma glucose concentrations and becomes the sole source of glucose.

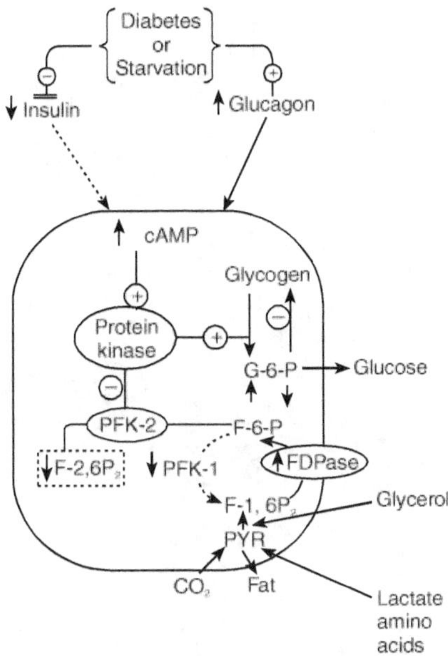

Figure 13.1   Enhancement of gluconeogenesis and glycogenolysis by glucagon in diabetes and starvation

As shown in Figure 13.1, both processes of gluconeogenesis and glycogenolysis are activated by increase in cyclic adenosine monophosphate (cAMP) in the hepatocyte. Phospho-fructokinase-1 (PFK-1) catalyses the formation of fructose 1,6 biphosphate (F-2, 6-P_2) in the glycolytic pathway, whereas PFK-2 synthesizes F-2, 6-P_2, a regulator of PFK-1 activity; cAMP-induced phosphorylation of the enzyme decreases PFK-1 and increases PFK-2. Decreased F-2, 6-P_2 decreases glycolysis and increases gluconeogenesis.

As shown in Figure 13.2 insulin decreases cAMP, deactivates protein kinase, and reverses changes in F-2, 6-P_2 and substrate flux over the glycolytic–gluconeogenic pathway produced by glucagon. Glycogen synthesis and lipogenesis are also increased.

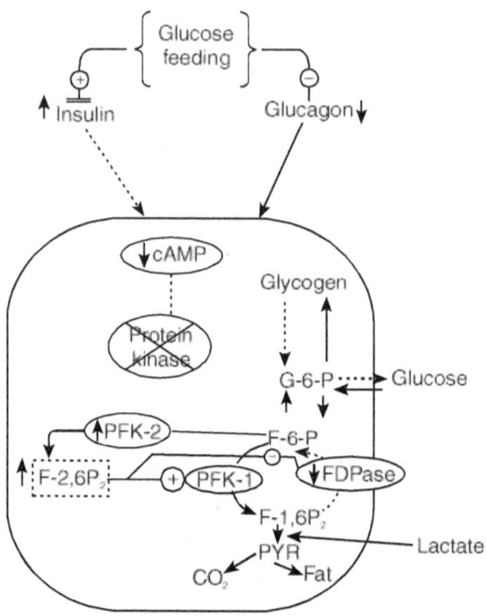

Figure 13.2 Inhibition of gluconeogenesis and activation of glycogen synthesis and lipogenesis by insulin

Exercise requires increased glucose production to counterbalance the increased glucose use that occurs with muscular work. Approximately 60 to 70% of the increased glucose production is mediated by increased glucagon secretion coupled with inhibited insulin release, another 30 to 40% is the result of epinephrine secretion. The changes in glucagon and insulin are associated with decreased fructose 2,6-diphosphate concentration in liver with resultant increased gluconeogenesis; simultaneous increase in epinephrine concentration produced increased concentration of fructose 2,6-diphosphate in non-exercising muscle with resultant stimulation of glycolysis and lactate production for use in gluconeogenesis. Combined glucagon deficiency and adrenergic blockade during exercise at 60% of maximal oxygen consumption produce profound hypoglycaemia between 30 and 60 minutes of the exercise bout.

Lipid Metabolism

Insulin and glucagon also play important roles in lipid metabolism; increased insulin concentrations stimulate lipogenesis and lipid storage, and the decreased insulin and increased glucagon levels seen in fasting promote lipolysis and lipid oxidation. The major function of stored triglyceride in adipose tissue is to act as an efficient energy reserve. Triglyceride stores can serve as a fuel to support many weeks of fasting, whereas stored carbohydrate is able to support a fast lasting only several hours. Stored triglycerides yield over two times as many calories per gram as either carbohydrate or protein, and require less than one half the intracellular water for storage.

In the fed state, insulin and glucose are required for lipogenesis. Glucose use is needed for fatty acid synthesis and esterification and supplies the following: (1) acetyl-CoA as a precursor of long-chain fatty acids, (2) α-glycerophosphate for esterification to fatty acids to form triglycerides and (3) nicotinamide adenine dinucleotide phosphate (NADPH). Insulin stimulates carrier-mediated glucose transport, activates pyruvate dehydrogenase for conversion of glucose to acetyl-CoA and inhibits lipolysis, thereby reducing palmitoyl-CoA, an inhibitor of lipogenesis.

Fatty acids stored in adipose tissue as triglycerides are derived from either dietary (chylomicrons) or endogenous (hepatic very low-density lipoproteins [VLDL]) sources. Preformed triglycerides are transported from the gastrointestinal tract and liver to adipose tissue, where they are hydrolysed by the enzyme lipoprotein lipase (LPL) on the cell surface of the capillary endothelium. Insulin has an important role in maintaining and stimulating the activity of lipoprotein lipase. In addition, insulin has a direct stimulatory effect on free fatty acid uptake by adipose tissue. During insulin deficiency, lipoprotein lipase activity is reduced and uptake of free fatty acids by adipose tissue is diminished.

In humans liver as well as adipose tissue are major sites of lipid synthesis, occurring when dietary fat is replaced by carbohydrate. The liver removes a large proportion of circulating free fatty acids delivered from adipose tissue in a concentration-dependent manner. Fatty acids synthesized in the liver are converted mainly to VLDL, which are secreted into plasma and then cleared from the circulation within minutes to hours by mechanisms similar to those involved in the removal of chylomicron triglycerides. During insulin deficiency, hexose monophosphate shunt activity is impaired, and NADPH is not provided for fatty acid synthesis. In addition, decreased glucose use reduces the availability of acetyl-CoA and citrate, which retards lipogenesis.

Lipolysis, with a net release of free fatty acids and glycerol from adipose tissue, occurs during periods of fasting, exercise, stress and uncontrolled diabetes mellitus. Low levels of insulin and increased glucagon concentrations enhance this mobilization of lipid from adipose tissue. Several hormones, including glucagon, catecholamines, thyroid stimulating hormone (TSH), and adrenocorticotrophic hormone (ACTH), play important roles in lipolysis, through cyclic-AMP-mediated stimulation of "hormone-sensitive lipase".

The insulin–glucagon ratio appears to be critical in regulating the hepatic metabolism of free fatty acids.

Activated fatty acids must be transported into the mitochondria for oxidation or conversion to ketone bodies, and neither free fatty acids nor their CoA derivatives can penetrate the inner mitochondrial membrane.

Carnitine palmitoyl transferase I is an enzyme present on the inner mitochondrial membrane, which reversibly transfers fatty acyl groups from CoA to carnitine and allows entry into the mitochondria. A second enzyme, carnitine palmitoyl transferase II irreversibly transfers the fatty acyl groups to mitochondrial CoA allowing them to undergo either β-oxidation or conversion to ketone bodies like aceto acetate and β-hydroxybutyrate. The activity of the key enzyme, carnitine palmitoyl transferase I is regulated through the effects of insulin and glucagon on malonyl-CoA concentrations. In addition to their effects on carnitine palmitoyl transferase I activity, low insulin and high glucagon concentrations also contribute to increased lipid oxidation and ketogenesis by increasing adipose tissue lipolysis and free fatty acid delivery.

Protein Metabolism

Insulin and other hormones play an important role in protein metabolism. In as small a period as several hours of starvation, protein catabolism is increased to provide amino acids for gluconeogenesis. With more prolonged starvation, metabolic adjustments occur that spare muscle protein, such as increased reliance by the central nervous system on ketone bodies as oxidative fuel. Nonetheless, muscle continues to yield a net release of amino acids as plasma insulin levels fall during prolonged starvation, the phenomenon can be reversed completely when insulin delivery is increased. When fuel supplier become plentiful, protein synthesis is restored.

Insulin lowers blood concentrations of several amino acids in normal as well as in diabetic subjects.

Prior exercise may potentiate the ability of insulin to increase amino acid uptake by muscle tissues. Exercise additionally inhibits amino acid release from skeletal muscle protein.

The effects of glucagon on amino acid metabolism are threefold:

1. to increase membrane transport of amino acids

2. to decrease protein synthesis and increase catabolism and

3. to increase amino acid conversion into glucose (gluconeogenesis)

THYROID HORMONES

Carbohydrate Metabolism

Thyroid hormones exert multiple effects on carbohydrate metabolism. Patients with hyperthyroidism frequently (30–50%) display mild to moderate degrees of glucose tolerance. Part of this abnormality is due to more rapid gastric emptying and intestinal absorption of glucose in hyperparathyroidism, whereas glucose absorption is delayed in hypothyroidism.

Hepatic glycogen stores are reduced in states of thyroid hormone excess. The glycogen content of liver and muscle tissues is decreased in hypothyroidism possibly reflecting a new balance between simultaneously decreased rates of glycogen synthesis and degradation.

Total body glucose turnover rates are increased in thyrotoxicosis. A major fraction of the increased glucose turnover is accounted for by increase in glucose recycling. Recycling through both the Cori (glucose–lactate) and the glucose–alanine cycles is increased in hyperthyroidism and decreased in hypothyroidism. Thyroid hormone facilitates these cycling phenomena by stimulating glucose transporter gene expression and enhancing glucose transport across the plasma membrane in muscle and liver cells.

Lipid Metabolism

Thyroid hormones also have an impact on multiple aspects of lipid metabolism, including lipid synthesis, mobilization and degradation. Degradation is affected more than synthesis, so the net effect of excess thyroid hormone is a decrease in total body lipid stores and plasma concentrations. Thyroid hormones increase lipolysis in adipose tissue by both directly stimulating cyclic AMP production and increasing the sensitivity to other lipolytic agents (catecholamines, TSH, ACTH, growth hormone, glucocorticoids and glucagon). Conversely, lipolysis is impaired in hypothyroidism.

The delivery of free fatty acids to peripheral tissues and the liver is increased in hyperthyroidism. Free fatty acid turnover rates are increased approximately twofold in thyrotoxicosis. Lipid oxidation rates also increase in thyrotoxicosis.

At a cellular level, lipid transfer, into and out of membranes is altered by thyroid hormone. Cholesterol is transferred out of erythrocyte plasma membranes into plasma in hyperthyroidism.

Protein Metabolism

With regard to protein metabolism, short-term administration of thyroid hormones produces increased liver protein and RNA content with a concomitant decrease in muscle protein. Thyroid hormone increases amino acid uptake by liver and increases incorporation of amino acids into protein. However, a very great excess of thyroid hormone has the opposite effect, i.e., with suppressed rates of protein synthesis and increased catabolism of collagen.

Nitrogen excretion is increased in thyrotoxicosis and nitrogen balance may be normal or negative depending on the intake.

In hypothyroidism, rates of protein synthesis and degradation are both decreased. Patients are usually in positive nitrogen balance.

GLUCOCORTICOIDS

Carbohydrate Metabolism

Corticosteroids in excess are known to produce increase in plasma insulin and glucose concentrations, i.e., a state of insulin resistance.

Glucocorticoids counteract the effects of insulin at numerous steps in glucose homoeostasis. First, rates of gluconeogenesis may be augmented by several mechanisms:

1. increased release of gluconeogenic precursors, i.e., amino acids and lactate from peripheral tissues,

2. increased activity of key gluconeogenic enzymes including pyruvate carboxylase and phosphoenol pyruvate carboxykinase, the unidirectional rate-limiting enzyme in the initiation of the gluconeogenic cascade from pyruvate and

3. stimulation of glucagon secretion by pancreatic alpha cells.

The second way in which glucocorticoid may affect glucose tolerance is by decreasing production of glucose transporters and promoting their sequestration in intracellular pools rather than in the plasma membrane.

Third, glucocorticoids decrease insulin binding to its receptor through decrease in receptor affinity.

Glucocorticoids stimulate hepatic glycogen deposition. The carbon source for this new liver glycogen is derived from breakdown of muscle protein with release of amino acids.

Lipid Metabolism

Glucocorticoids appear to exert a permissive effect on lipolysis through activation of cyclic-AMP dependent hormone-sensitive lipase in the adipocyte.

Prolonged treatment with glucocorticoids may result in increased plasma triglyceride concentrations. LDL uptake is also impaired. Chronic glucocorticoid treatment may result in a "fatty liver" because of increased lipolysis and free fatty acid delivery associated with enhanced uptake of free fatty acids by the hepatic cell.

Protein Metabolism

One of the major metabolism effects of glucorticoids is to stimulate skeletal muscle protein breakdown. Many of the clinical features of Cushing's syndrome, such as the loss of bone density, increased capillary fragility and dermal atrophy, muscle wasting and growth retardation in children are attributable to this augmented proteolysis. In addition to increasing protein breakdown, corticosteroids also inhibit incorporation of amino acids into muscle protein.

GROWTH HORMONE

Growth hormone has direct catabolic effects on glucose and lipid metabolism.

Carbohydrate Metabolism

Growth hormone lowers plasma glucose levels by directly stimulating β-cell insulin secretion and also by stimulating glucose use in peripheral tissues.

Glucose intolerance in association with acromegaly has been reported in 60–70% of the patients. There is also hyperinsulinaemia and resistance to insulin that occur in acromegaly. A chronic excess of growth hormone results in increased rates of hepatic glucose production and decreased glucose use by peripheral tissues, which is largely due to decreased glucose oxidation. Decreased peripheral glucose use in acromegaly is associated with a decreased number of insulin receptors on peripheral blood monocytes.

Chronic growth hormone deficiency produces increased insulin sensitivity.

Lipid Metabolism

Free fatty acid levels and ketone body levels are increased when excessive growth hormone is produced as rate of lipolysis increases. Fatty acid synthesis is inhibited by growth hormone.

Protein Metabolism

One of the major effects of growth hormone is the promotion of linear growth and skeletal maturation. There is an increased synthesis of DNA, RNA and protein. There is a negative nitrogen balance and protein and RNA contents decrease in various tissues during growth hormone deficiency.

CATECHOLAMINES

These are epinephrine, norepinephrine and dopamine. Norepiphrine is secreted from the sympathetic neurons throughout the body. The principal secretary product of the adrenal medulla is epinephrine. The major stimuli to sympathetic nervous system stimulation and adrenomedullary secretion are physical exercise, circulatory dysfunction, trauma, cold exposure, pain, emotional stress and hypoglycaemia.

Carbohydrate Metabolism

Catecholamines have multiple effects on carbohydrate metabolism.

Catecholamines increase glycogen breakdown in liver and muscle tissues causing hepatic glycogenolysis.

The effects of catecholamines on muscle glycogen metabolism are antagonized by insulin and depend on glucocorticoids.

Lipid Metabolism

The major effect of catecholamines on lipid metabolism is augmentation of lipolysis. Both epinephrine and norepinephrine activate hormone-sensitive lipase in adipose tissue, liver, heart and skeletal muscle.

Catecholamines increase lipogenesis and ketogenesis. Hepatic triglyceride synthesis is increased. Catecholamines also increase plasma lipid levels as also cholesterol synthesis goes up.

Protein Metabolism

Catecholamines have insulin-like effects on plasma amino acid levels.

In contrast to insulin, epinephrine increases uptake of gluconeogenic amino acids.

SEX STEROIDS AND PROLACTIN

Carbohydrate Metabolism

Oestrogen therapy exacerbates mild Diabetes mellitus. Progesterone found in oral contraceptive pills also alters glucose tolerance.

Lipid Metabolism

Plasma levels of VLDL cholesterol are increased because of enhanced hepatic production in oestrogen therapy. In contract, the clearance of LDL is enhanced by estrogens, in part because of increased hepatic excretion of cholesterol in the bile. Oestrogens also produce significant increases in plasma HDL-cholesterol levels.

Protein Metabolism

Testosterone augments skeletal muscle mass. Prolactin has an effect similar to growth hormone.

REVIEW QUESTIONS

1. Discuss the role of glucagon in diabetes.

2. Discuss the role of hormones in lipid metabolism.

CRITICAL THINKING QUESTION

1. Is insulin the only hormone that increases uptake of glucose by cells?

14

IMMUNOLOGY AND NUTRITION

Did You Know?

- There are three major groups of phagocytic cells 1) neutrophils 2) the cells of the mononuclear phagocyte system including monocytes and macrophages 3) organ-specific phagocytes in the liver, spleen, lymph nodes, etc.

- Leucocytes are attracted to the site of an infection by a process called chemotaxis.

- The attractants are chemical in nature known as chemokines.

- Antigens are molecules that stimulate the production of specific antibodies.

- Bacteria can enter a break in the skin.

- Antibody proteins are also called as immunoglobulins.

- There are five subclasses of antibody or immunoglobulins IgG, IgA, IgM, IgD and IgE.

- The thymus processes T lymphocytes and secretes hormones that are required for an effective immune response.

Immunology is the study of the immunity of living organisms to harmful agents. Mammals produce specifically reacting glycoproteins against foreign substances. These protective glycoproteins are called antibodies because they react and neutralize the "foreign bodies", which induce the production of the specific antibodies. The foreign molecules capable of inducing the formation of antibodies were called antigens because they generated antibody in the host.

Serum antibodies are special glycoproteins or globulins, and are concerned with immunity and hence called immunoglobulins. Five classes of immunoglobulins are recognized, namely, IgG, IgM, IgA, IgE and IgD. The production of antibodies by a host is commonly referred to as the humoral immune response.

B lymphocytes are the cells responsible for antibody formation. In addition, another class of lymphocytes, designated T lymphocytes, can mediate cell-mediated immunity (CMI) directly as cytotoxic immune cells or through macrophages that are mobilized and activated by T-lymphocyte products (lymphokines). Lymphokines are cytokines or mediator products that act on many cell types. The origin of cells of the immune system is given in Figure 14.1.

CELLS OF THE IMMUNE SYSTEM

The immune system is divided into the central lymphoid system and the peripheral lymphoid system.

The central lymphoid system comprises the bone marrow, thymus and the bursa equivalent whereas the peripheral system comprises the spleen, lymph nodes, tonsils, mucous membrane associated lymphoid tissue and other diffuse lymphoid tissues of the body.

Immune responses are mediated by numerous cell types and complex cell interactions. These cells include lymphocytes, macrophages, granulocytes, mast cells and dendritic cells. The origin of cells of the immune system is given in Figure 14.1.

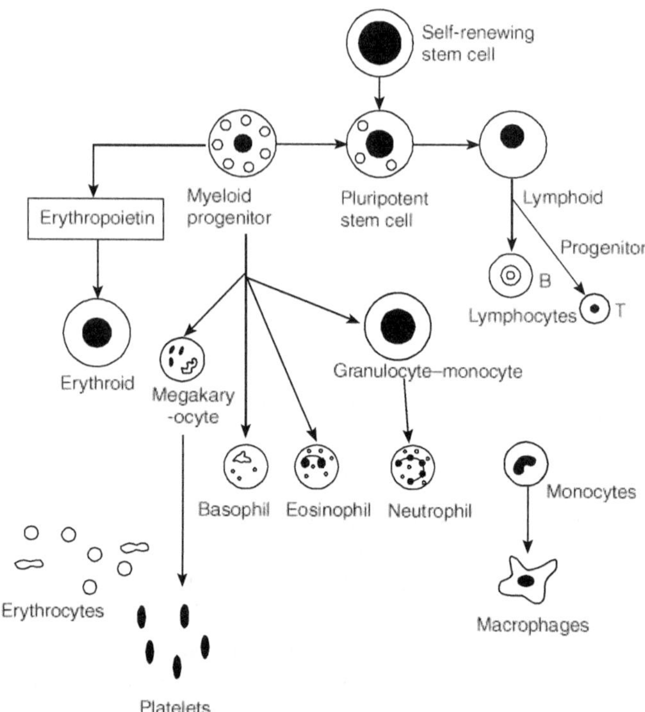

Figure 14.1   Origin of cells of the immune system

Granulocytes

Blood granulocytes, which comprise neutrophils, eosinophils and basophils are found in the blood as fully differentiated and short-lived cells 12 to 15 μm in diameter.

Neutrophils comprise about 75% of the circulating blood leucocytes. Neutrophils provide the first line of phagocytic defense against many microorganisms. They liberate the fever-producing agent referred to as endogenous pyrogen, and are active in the digestion of foreign material and dead tissue. The human adult produces 126 million neutrophils every day.

Eosinophils comprise about 2 to 6% of the circulating leucocytes. They can be important effector cells in immunity against several animal parasitic infections.

The basophil normally comprises about 0.05 to 1% of the circulating leucocytes. It possesses receptors that bind to the Fc segment of the IgE antibody molecule. The mast cell which is the functional counterpart of the basophil in tissue, is abundant in perivascular and peribronchiolar connective tissue adjacent to smooth muscle. Basophils and mast cells become activated and release mediators when IgE molecules are bound to the Fc receptors and are cross-linked by reacting with specific antigens. Mast cell rupture results from tissue trauma, and can also result in the release of mediators that can cause significant inflammation.

Macrophages

Macrophages are large, long-lived mononuclear phagocytic cells that range in diameter from about 12 to 25 μm. Macrophage precursors (promonocytes) pass continually from the bone marrow to the blood as small (12 to 15 μm) non-activated cells. They comprise about 5% of the blood leucocytes. Many of the monocytes in the blood become fixed to vessel walls within organs such as the liver and the spleen, where they engage in phagocytic and pinocytic activity, which in turn induces their activation and maturation. Some monocytes migrate through the tight junctions of vessel walls to reach extravascular sites and tissues such as body cavities, lung alveoli, lymphoid organs and even the brain. Antibodies can bind to macrophages, which enables the macrophage to adhere and attack the membranes of foreign mammalian cells containing the corresponding antigens.

Lymphocytes

Lymphocytes play the central role in the specific immune response. A normal human adult possesses approximately 10^{12} lymphocytes most of which reside in the lymphoid organ systems.

The stem cell progenitor of all lymphocytes resides in bone marrow.

The three major groups of lymphocytes are B cells, T cells and null cells. The latter group lacks most of the surface characteristics of either B or T cells. Null cells appear to play an important role as killer cells in anti-tissue immunity. The overall specific immune response is divided into two major types of response: the humoral and the cell-mediated response. The immune cell responsible for the humoral antibody response is the B cell, whereas the T cell is responsible for mediating cell-mediated immune response.

B cells are responsible for the production of all the immunoglobulins. However, certain antigens, referred to as thymus-dependent (TD) antigens, can induce antibody formation by B cells only if T helper cells are present.

Plasma cells are the end cells of antigen-stimulated B cells. These are highly specialized, fully differentiated cells derived from B cells that produce and secrete immunoglobulin. Plasma cells are found only in extravascular sites and do not circulate in the vascular compartment. The plasma cells are characteristic because of enormous amounts of dilated endoplasmic reticulum in the cytoplasm.

EFFECTS OF MALNUTRITION ON IMMUNE RESPONSE

Many studies have shown that severe protein deficiencies suppress antibody formation. In addition, deficiencies of pyridoxine, pantothenic acid and pteroyl glutamic acid also result in suppressed antibody responses.

Effect of Protein and Protein-energy Malnutrition

The number of plaque-forming lymphocytes that become activated and the corresponding amount of antibody synthesized was directly correlated with protein or protein-energy intake.

Studies have shown that on a 3% protein intake, tumour cell killing was reduced in mice. Test groups (guinea pigs) fed on a 5% protein diet showed a decline in the phagocytic function.

Tuberculin reactivity disappeared on a 0.5% protein diet.

Severe-protein-energy malnutrition in humans results in impairment of both humoral or cell-mediated immune functions. Thymic atrophy and T-cell deficiencies are common in undernourished children. A depression of T-helper cells and possibly an increase in T-suppressor cells also can occur in PEM. There is a decreased level of secretory IgA in tears, saliva, and pharyngeal secretions that could be responsible for the compromised resistance to organisms that cause respiratory infections (Table 14.1).

Table 14.1 Summary of effects of PEM on immune functions

Lymphoid	
Thymus	Decreased
Spleen	Decreased
Lymph nodes	Decreased
Total circulating lymphocytes	Decreased
Humoral immunity	
Circulating B-lymphocytes	Decreased or normal
Serum Ig levels	Decreased
Cellular immunity	
Circulating T-lymphocytes	Decreased
Tumour cytotoxicity	Decreased
Phagocytic function	
Monocyte chemotaxis	Decreased
Reticuloendothelial system function	Decreased
Intracellular killing	Decreased or normal

SINGLE NUTRIENT DEFICIENCIES

Many animal studies have been helpful in arriving at a consensus of how single nutrients affect the immune system.

Amino Acids

The deficiencies of essential amino acids can result in impairment of the humoral response. Chronic deficiencies of branched-chain or sulphur-containing amino acids result in a depletion of cells from lymphoid tissues. Rats with tryptophan-deficient diets exhibit depressed IgG and IgM. Tryptophan deficiency also resulted in depressed cytotoxic function, with further reduction with leucine deficiency at the 25% dietary level. Depression of serum blocking and haemagglutinating antibody levels was noted with single deficiency of methionine-cysteine, valine, tryptophan threonine and phenyl alanine-tyrosine reduced to the 25% dietary level.

Zinc

Zinc deficiency causes atrophy of lymphoid tissue and produces abnormalities in both cellular and humoral immunity.

Zinc deficiency may be one of the most prevalent nutritional problems worldwide. The average adult must obtain about 15 mg of zinc per day from the diet. Because the body stores of this element are limited, a constant steady-state intake of zinc is required. Zinc deficiency is caused by diets that contain large amounts of grains, cereals, and unleavened bread, and in countries where little animal products are eaten. Because these foods contain high amounts of phytic acid which is a zinc chelator and hence prevents zinc absorption.

Zinc is a cofactor for at least 90 metalloenzymes, including enzymes required for transcription and translation. Because cells of the immune system are under continuous proliferation and differentiation and require numeral enzymes that use zinc as a cofactor, this deficiency affects both the affector and effector responses. T-cell maturation and replication are absolute requirements for normal function.

Lymphopenia, retarded wound healing, thymic atrophy, reduced capacity to exhibit delayed hypersensitivity, and increased susceptibility to disease are seen in zinc-deficient children.

In zinc deficiency, there is reduction in IgM and IgG plaque-forming cells, impaired secondary antibody responses reduced responses, of lymphocytes, impaired cell-mediated immunity and impaired T-cell maturation.

Zinc deficiency also can lead to a breakdown in the killer T-cell function and helper T-cell function.

Iron

Human subjects with iron deficiency exhibit impaired delayed hypersensitivity reactions as well as defective neutrophil and macrophage killing functions.

Iron deficiency results in atrophy of lymphoid tissues and depletion of lymphocytes with a subsequent decrease in antibody production. Infants receiving adequate iron and vitamins experience only half the incidence of respiratory infections.

Iron also plays a critical role in neutrophils, which use iron-dependent enzymes and proteins like myeloperoxidase in its killing action of phagocytozed bacteria.

Iron overload also appears to impair antigen-specific immune responses and reduces the number of helper precursor cells, i.e., results in impaired generation of cytotoxic T cells, enhanced suppressor T-cell activity and reduced proliferative capacity of helper T cells.

Manganese

Manganese may stimulate macrophage spreading on antigen-antibody coated surfaces. An excess of manganese can cause elevated antibody titres and other non-specific resistance factors.

Copper

Deficiencies of copper appear to prevent the reticuloendothelial system from reacting normally to infection. There was an inhibition of T-cell mitogen responses, especially T-helper activity in copper deficiency. It also causes an increase in infections controlled by T-cell-mediated immunity. It also impairs the maturation and formation of erythrocytes and neutrophils. Copper is also required for lymphocyte development.

Magnesium

IgG levels were low during magnesium deficiency. Rats when fed a magnesium-deficient diet developed persistent leucocytosis with increased numbers of neutrophils, mononuclear cells and eosinophils.

Selenium

Selenium is an important component of the selenoprotein enzyme glutathione peroxidase, which probably is the major site for its function. This enzyme catalyses the reduction of organic and inorganic hydroperoxides using glutathione as the electron donor, which in turn functions as a potent antioxidant. Because neutrophils and macrophages generate large amounts of reactive oxygen species during an oxidative burst, it is important for the cell to neutralize an excess of these oxidants. It is possible that a selenium deficiency could result in a dysfunction of this peroxidase, reducing the ability of cells to protect themselves against the damage by oxidants generated during the burst. Therefore selenium appears to function as an important antioxidation in the metabolism of phagocytes which allows them to control autotoxic damage.

Other Minerals

Minerals like cadmium, lead, chromium, mercury, nickel, vanadium and gold are not required but are of potential importance because of their toxicity for cells and their potential deleterious effects on the immune system.

EFFECTS OF THE B-COMPLEX VITAMINS ON THE IMMUNE RESPONSE

Deficiencies of the B-group vitamins results in lymphopenia and atrophy of peripheral lymphoid tissue. Deficiencies of pyridoxine, pantothenic acid, riboflavin folate and vitamin B_{12} have the greatest effect, whereas biotin, thiamine and folic acid have lesser effects.

Pyridoxine

B_6 deficiency causes atrophy of lymphoid tissue, depressed primary and secondary antibody responses, reduced mixed lymphocyte responses and decreased dermal hypersensitivity.

Pantothenic Acid and Riboflavin

There is impaired antibody response, and a deficiency of riboflavin causes a decrease in thymic effect.

Folic Acid

In folic acid deficiency there is impaired neutrophil function, impaired responses of lymphocytes, impaired cytotoxic T-cell function and an impaired antibody response.

An impairment of phagocytosis by neutrophils has also been reported. Impairment of antibody response to diphtheria toxoid and human red blood cells has been reported to occur in rats with folate deficiency.

Vitamin B_{12}

Vitamin B_{12} is essential for nucleic acid synthesis, which is needed for cell growth and division. Low levels of B_{12} can also result in decreased transport and use of folic acid. There is significant depression of lymphocyte transformation to phytohaemagglutinin in patients with pernicious anaemia. Neutrophils from pernicious anaemic patients had an impaired ability to phagocytoze and kill *Staphylococcus aureus*.

Biotin

Biotin deficiency has been clearly associated with impaired humoral and cell-mediated responses. Biotin is a coenzyme for several enzyme-mediated carboxylation reactions required for fat and carbohydrate metabolism. Rats fed a biotin-deficient diet exhibited a significant reduction in the size of the thymus.

Thiamine

Vitamin B_1 functions as a coenzyme in the decarboxylation reactions resulting in energy derived from glucose. As a consequence of a thiamine deficiency, increased levels of pyruvic acid and impairment of erythrocyte transketolase activity occurs.

Studies report that thiamine deficiency results in an increased susceptibility to *Salmonella typhimurium* in mice. Other reports suggest that thiamine reverses inhibition of neutrophil migration.

Vitamin C

Vitamin C deficiency results in impaired inflammatory responses, impaired cytotoxic T-cell activity, impaired phagocytic function of neutrophils and macrophages.

Vitamin A

Vitamin A plays a major role in maintaining the functional integrity of epithelial and mucosal surfaces and promotes the production of mucous secretions as well as salivary and sweat gland lysozyme.

Vitamin A deficiency causes a reduction in lymphocyte response to infection. There is also an increased susceptibility to colon and liver cancer in vitamin A deficiency.

Macrophage function has been enhanced by high levels of vitamin A to produce increased amounts of IL-1.

Vitamin E

In vitamin E deficiency there is impaired antibody responses, impaired responses of lymphocytes to mitogens and impaired phagocytic function.

Immunological Reactions in Malnourished Children

One of the early changes noted in severely malnourished children is thymic atrophy. Post-mortem studies on African children with marasmus and Kwashiorkor revealed greatly reduced weights of thymus, spleen, Peyer's patches and appendices. In particular, paracortical areas were depleted and germinal centres were absent.

Children with moderate and severe protein malnutrition experience significant suppression (about 50%) of lysozyme secretion into tears. There is also a reduction in the production of IgA.

FATTY ACIDS AND IMMUNE SYSTEM

- Fatty acids function as components of membranes and as precursors for hormone-like compounds termed eicosanoids.
- Essential fatty acid deficiency impairs cell-mediated immunity.
- Cell-mediated immunity and natural killer cell activity decrease as the fat content of the diet increases.
- Within a high-fat diet, different fatty acids can exert different effects.
- Increasing linoleic acid intake decreases cell-mediated immunity.
- Fish oil has been shown to improve symptoms in rheumatoid arthritis.

PROBIOTICS

Probiotics is the name given to desirable bacteria that colonize the gut. They contribute to the immunological protection of the host by creating a barrier against colonization by pathogenic bacteria.

Probiotic organisms are found in fermented foods. The organisms include lactic acid bacteria. In addition to creating a barrier effect, some of the metabolic products of probiotic bacteria (a class of antiobiotic proteins termed bacteriocins) may inhibit growth of the pathogenic organisms. Probiotic bacteria also compete with pathogens for nutrients and prevent growth of pathogens.

REVIEW QUESTIONS

1. Write a note on the cells of the immune system.

2. Discuss the effects of PEM on immune function.

3. Discuss single nutrient deficiencies and their effect on the immune system.

CRITICAL THINKING QUESTION

1. Why do deficiencies affect the immunity?

15

SPORTS NUTRITION

Did You Know?

- Endurance training results in an improved ability to obtain ATP from oxidative phosphorylation.

- Increased exercise increases the size and number of mitochondria.

- Maximal oxygen uptake, obtained during very strenuous exercise, averages 50 ml of O_2 per minute per kilogram body weight in males.

- Athletes produce less lactic acid at a given level of exercise.

- The greater the level of physical training the higher the proportion of energy derived from the oxidation of fatty acids during exercise.

- Athletes are less subject to fatigue.

- New ATP can be quickly produced from the combination of ADP with phosphate derived from phosphocreatine.

- Endurance training causes hypertrophy of muscle fibres because of an increase in the size and number of myofibrils.

There are no other normal stresses to which the body is exposed that even nearly approaches the extreme stresses of heavy exercise. In fact, if some of the extremes of exercise were continued for even slightly prolonged periods of time, they might easily be lethal. To give a simple example: in a person who has extremely high fever, the body metabolism increases to about 100% above normal. By comparison, the metabolism of the body during a marathon race increases to 200% above normal.

THE MUSCLES IN EXERCISE

The strength of a muscle is determined mainly by its size, with a maximum contractile force between 3 kg and 4 kg per square centimetre of muscle cross-sectional area. Hence, the more the cross-sectional area the more will be the muscle strength. The muscle strength (cross-section) can be increased with testosterone (in men) and also through an exercise training programme which increases muscle development and increased muscle strength. To give an example of muscle strength, a world-class weight lifter might have a quadriceps muscle with a cross-sectional area as great as 150 sq. cms. This would be equal to a maximum contractile strength of 525 kgs (1155 lbs) with all this force applied to the patellar tendon. Therefore, one can readily understand how it is possible for this tendon to be ruptured or to be avulsed from its insertion into the tibia below the knee. Also when such forces occur in tendons that span a joint, similar forces are applied as well to the surfaces of the joints or sometimes to ligaments spanning the joints thus accounting for such happenings as displaced cartilages, compression fractures about the joint, or torn ligaments.

The holding strength of muscles is about 40% greater than the contractile strength, i.e., if a muscle is already contracted, and a force then attempts to stretch out the muscle, as occurs when landing after a jump, this requires about 40% more force than can be achieved by a shortening contraction. Therefore, the force of 525 kg calculated previously for the patellar tendon becomes 735 kg. This obviously further compounds the problems of the tendons, joints and ligaments. It can also lead to internal tearing in the muscle itself.

The power of muscle contraction is different from muscle strength, for power is a measure of the total amount of work that the muscle can perform in a given period of time. This is determined not only by the strength of muscle contraction but also by its distance of contraction and the number of times that it contracts each minute. Muscle power is generally measured in kilogram metres per minute. That is, a muscle that can lift a kilogram weight to a height of 1 metre or that can move some object laterally against a force of 1 kilogram for a distance of a metre in 1 minute is said to have a power of 1kg-m/min.

The final measure of muscle performance is endurance. This, to a great extent, depends on the nutritive support for the muscle—more than anything else on the amount of glycogen that has been stored in the muscle prior to the period of exercise. A person on a high-carbohydrate diet stores far more glycogen in muscles than a person on either a mixed diet or a high-fat diet. Therefore, endurance is greatly enhanced by a high-carbohydrate diet. When athletes run at speeds typical for a marathon race, their endurance as measured by the time that they can sustain the race until complete exhaustion, is approximately the following.

Type of diet	Minutes
High-carbohydrate diet	240
Mixed diet	120
High-fat diet	85

The corresponding amounts of glycogen stored in the muscle are approximately the following:

Type of diet	g/kg muscle
High-carbohydrate diet	40
Mixed diet	20
High-fat diet	6

MUSCLE METABOLIC SYSTEMS IN EXERCISE

The metabolic systems important in understanding the limits of physical activity are

1. The phosphagen system
2. The glycogen–lactic acid system
3. The aerobic system

The Phosphagen System

The basic source of energy for muscle contraction is adenosine triphosphate (ATP). ATP molecules have high-energy phosphate bonds. Each of these bonds stores 7300 calories of energy per mole of ATP under standard conditions. Therefore when one phosphate radical is removed from the molecule, 7300 calories of energy that can be used to energize the muscle contraction process is released. Then when the second phosphate radical is removed, still another 7300 calories becomes available.

The amount of ATP present in a muscle even in the well-trained athlete, is sufficient to sustain maximal muscle power for only about 3 seconds. Hence, new ATP should be formed continuously even during the performance of athletic events.

Release of energy from creatine phosphate Creatine phosphate is another chemical compound that has a high-energy phosphate bond. This can decompose to release phosphate ion and releases large amounts of energy. The high-energy phosphate bond of creatine phosphate releases 10,300 calories per mole, in comparison with 7300 of ATP.

Muscle cells have two to four times as much creatine phosphate as ATP. Energy is transferred from creatine phosphate to ATP within a small fraction of a second. Therefore all the energy stored in the muscle is instantaneously available for muscle contraction.

The cell creatine phosphate plus its ATP is called the phosphagen energy system. These together can provide maximal muscle power for a period of 8 to 10 seconds. Thus energy from the phosphagen system is used for maximal short bursts of muscle power (Figure 15.1).

The Glycogen–Lactic Acid System

The stored glycogen in muscle can be split into glucose and the glucose then utilized for energy. Glucose during glycolysis splits into two pyruvic acid molecules, and energy is released to form several ATP molecules. The pyruvic acid then enters the mitochondria of the muscle cells and reacts with oxygen to form still many more ATP molecules. However, when there is insufficient oxygen for this second stage of glucose metabolism to occur, most of the pyruvic acid is converted into lactic acid, which then diffuses out of the muscle cells into the interstitial fluids and blood. Therefore, in effect, much of the muscle glycogen becomes lactic acid, but in doing so considerable amounts of adenosine triphosphate are formed entirely without the consumption of oxygen (Figure 15.1).

Another characteristic of the glycogen–lactic acid system is that it can form ATP molecules about 2.5 times as rapidly as can the oxidative mechanism of the mitochondria. Therefore, when large amounts of ATP are required for short to moderate periods of muscle contraction, this anaerobic mechanism is used as a source of energy. However, it is not as rapid as the phosphagen system but only about half as rapid. Under optimal conditions the glycogen–lactic acid system can provide 1.3 to 1.6 minutes of maximal muscle activity in addition to the 8 to 10 seconds provided by the phosphagen system.

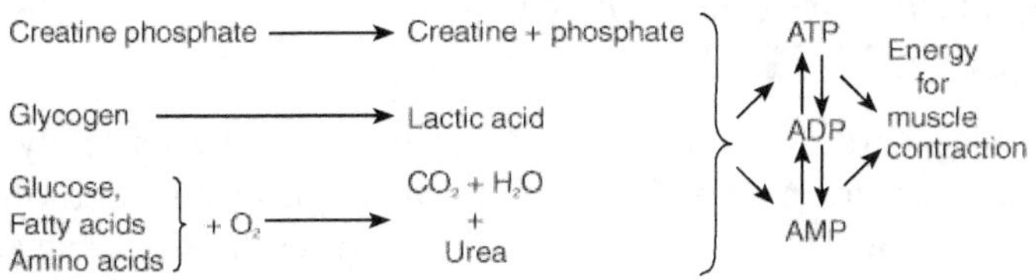

Figure 15.1 Release of energy anaerobically

The Aerobic System

The aerobic refers to the oxidation of foodstuff in the mitochondria to provide energy.

In comparing the aerobic system of energy supply with the glycogen–lactic acid system and the phosphagen system, the relative maximum rates of power generation in terms of ATP utilization are the following.

System	M of ATP/min.
Aerobic system	1
Glycogen–lactic acid system	2.5
Phosphagen system	4

On the other hand when comparing the system for endurance, the relative values are the following.

System	Time
Phosphagen system	8 to 10 seconds
Glycogen–lactic acid system	1.3 to 1.6 minutes
Aerobic system	Unlimited time (as long as nutrients last)

Hence, the aerobic system is required for prolonged athletic activity.

NUTRIENTS USED DURING MUSCLE ACTIVITY

A high carbohydrate diet and a large store of muscle glycogen is necessary for maximal athletic performance as the carbohydrates are used for energy by preference. The muscles use large amounts of fat for energy in the form of fatty acids and aceto acetic acid, and also use to a much lesser extent proteins in the form of amino acids. In those endurance athletic events that last longer than 4 to 5 hours, the glycogen stores of the muscle become depleted and now the muscle depends upon energy from other sources, mainly from fats.

Most of the energy is derived from carbohydrate during the first few seconds or minutes of the exercise, but at the time of exhaustion, as much as 60 to 85% of the energy is being derived from fats, rather than carbohydrates. Glycogen stored in the liver is also released into the blood in the form of glucose and then can be taken up by the muscles as an energy source. In addition, glucose solutions given to an athlete to drink during the course of an athletic event (in optimal concentrations of 2 to 2.5%) can provide as much as 30 to 40% of the energy required during prolonged events such as marathon races.

Then, if muscle glycogen and blood glucose are available, these are the energy nutrients of choice for intense muscle activity. Even so, for a real endurance event one can expect fat to supply more than 50% of the required energy after about the first 3 to 4 hours.

BODY FLUIDS AND SALT IN EXERCISE

As much as a 2 to 5 kg weight loss has been recorded in athletes in a period of 1 hour during endurance athletic events under hot and humid conditions. Essentially all of this weight loss results from loss of sweat. Loss of enough sweat to decrease body weight only by 3% can significantly diminish a person's performance and a 5 to 10% rapid decrease in weight can often be very serious, leading to muscle cramps, nausea and other effects. Therefore, it is essential to replace fluid as it is lost.

Replacement of Potassium Salts

Sweat contains a large amount of salt, and hence all athletes should take salt (sodium chloride) tablets when performing exercise on hot and humid days. Furthermore, if an athlete becomes acclimatized to the heat by progressive increase in athletic exposure over a period of 1 to 2 weeks rather than performing maximal athletic feats on the first day, the sweat glands also become acclimatized, so that the amount of salt lost in the sweat is only a small fraction of that prior to acclimatization. This sweat gland acclimatization results mainly from increased aldosterone secretion by the adrenal cortex. The aldosterone in turn has a direct effect on the sweat glands, increasing the reabsorption of sodium chloride from the sweat.

Potassium loss results from the increased secretion of aldosterone during heat acclimatization, which increases the loss of potassium in the urine as well as in the sweat. Hence, supplemental fluids for athletes contain properly proportioned amounts of potassium and sodium, usually in the form of fruit juices.

Carbohydrates and fat are always oxidized as a mixture, and whether carbohydrate or fat is the predominant fuel depends on a variety of factors.

Exercise intensity   At higher exercise intensities, more carbohydrate and less fat will be utilized. Carbohydrate can be utilized aerobically at rates up to about 4 g/min. The breakdown during very high intensity exercise can amount to 7 g/min.

Duration of exercise   Fat oxidation increases and carbohydrate oxidation decreases as the exercise duration increases. Typical fat oxidation rates are between 0.2 and 0.5 g/min. but values of over 1.0–1.5 g/min. have been reported after 6 hours of running. The contribution of fat to energy expenditure can even increase to as much as 90%. This increased fat oxidation is likely to be caused by a reduction in muscle glycogen stores towards the later stages of prolonged exercise.

Level of aerobic fitness   After endurance training, the capacity to oxidize fatty acids increases and fat oxidation at the same absolute and relative exercise intensity is higher.

Diet   Substrate utilization usually reflects the diet. A high carbohydrate diet will promote carbohydrate oxidation, whereas a low carbohydrate diet will reduce body carbohydrate stores and result in lower rates of carbohydrate oxidation.

Carbohydrate intake before or during exercise   After an overnight fast most of the energy requirement is covered by the oxidation of fatty acids, derived from adipose tissue. Lipolysis in adipose tissue is mostly dependent on the concentrations of hormones (epinephrine to stimulate lipolysis and insulin to inhibit lipolysis). At rest a large percentage (about 70%) of the fatty acids liberated after lipolysis will be re-esterified within the adipocyte and approximately 30% of the fatty acids will be released into the systemic circulation. Resting plasma fatty acid concentrations are typically between 0.2 and 0.4 mmol/l. As soon as exercise is initiated, the rate of lipolysis and the rate of fatty acid release from adipose tissue are increased. During moderate-intensity exercise, lipolysis increases approximately threefold, mainly because of an increased β-adrenergic stimulation. In addition, during moderate-intensity exercise the blood flow to adipose tissue is doubled and the rate of re-esterification is halved. Blood flow in skeletal muscle is increased dramatically and therefore the delivery of fatty acids to the muscle is increased.

FUELLING UP BEFORE EXERCISE

Carbohydrate stores in the muscle and liver should be well filled prior to exercise, particularly in the competition setting where the athletes want to perform at their best. The key factors in glycogen storage are dietary carbohydrate intake and in the case of muscle stores, tapered exercise or rest.

Pre-event Meal

Food and fluids consumed in the 4 hours before an event may help to achieve the following sports nutrition goals:

- To continue to fill muscle glycogen stores if they have not fully restored or loaded since the last exercise session.

- To restore liver glycogen levels, especially for events undertaken in the morning where liver stores are depleted from an overnight fast.

- To ensure that the athlete is well hydrated.

- To prevent hunger, yet avoid gastrointestinal discomfort and upset during exercise.

- To include foods and practices that are important to the athlete's psychology or superstitions.

Consuming carbohydrate-rich foods and drinks in the pre-event meal is especially important in situations where body carbohydrate stores have not been fully recovered and/or where the event is of sufficient duration and intensity to deplete these stores. The intake of a substantial amount of carbohydrate (200–300 g) in the 24 hour before exercise has been shown to enhance various measures of exercise performance compared with performance undertaken after an overnight fast.

However, carbohydrate intake before exercise may have negative consequences for performance, especially in the hour before exercise. Carbohydrate intake causes a rise in plasma insulin concentrations, which in turn suppresses the availability and oxidation of fat as an exercise fuel. The final result is an increased reliance on carbohydrate oxidation at the onset of exercise leading to faster depletion of muscle glycogen stores and a decline in plasma glucose concentration. This problem can be avoided or diminished by a number of strategies.

- ♂ Consume a substantial amount of carbohydrate (>75 g) rather than a small amount so that the additional carbohydrate more than compensates for the increased rate of carbohydrate oxidation during the exercise.

- ♂ Choose a carbohydrate-rich food or drink that produces a low glycaemic index response (i.e., a low blood glucose and insulin response) rather than a carbohydrate source that has a high glycaemic index.

- ♂ Consume carbohydrates throughout the exercise session.

FAT METABOLISM AND PERFORMANCE

At rest and during exercise skeletal muscle is the main site of oxidation of fatty acids. In resting conditions, and especially after fasting, fatty acids are the predominant fuel used by skeletal muscle. During low intensity exercise, metabolism is elevated several-fold compared with resting conditions and fat oxidation is increased. When the exercise intensity increases, fat oxidation increases further.

FAT OXIDATION AND DIET

Diet has marked effects on fat oxidation. In general, a high-carbohydrate, low-fat diet will reduce fat oxidation, whereas a high-fat, low-carbohydrate diet will increase fat oxidation.

FAT INTAKE DURING EXERCISE

Long-chain Triglycerides

Nutritional fats include triglycerides, phospholipids and cholesterol of which only triglycerides can contribute to any extent to energy provision during exercise. In contrast to carbohydrate nutritional fats reach the circulation only slowly as they inhibit gastric emptying. Digestion and absorption of fats are rather slow processes compared with the digestion and absorption of carbohydrates. Bile salts produced by the liver and lipase secreted by the pancreas are needed for the breakdown of long-chain triglycerides into glycerol and three long-chain fatty acids. The fatty acids are then transported in chylomicrons via the lymphatic system, which ultimately drains into the systemic circulation. Long-chain dietary fatty acids typically enter the blood 3–4 hours after ingestion. The rate of breakdown of chylomicron-bound triglycerides by muscle is relatively slow. The primary role of these triglycerides in chylomicrons is the replenishment of intramuscular fat stores after exercise. Therefore, it is advisable to reduce the intake of fat during exercise to a minimum.

Medium-chain Triglycerides

Medium-chain fatty acids contain 8–10 carbons while long-chain fatty acids have 12 or more carbons. Medium-chain triglycerides (MCT) are more polar and therefore more soluble in water and they are more rapidly digested and absorbed in the intestine. Furthermore, medium-chain fatty acids follow the portal venous system and enter the liver directly, while long-chain fatty acids are passed into the systemic circulation slowly via the lymphatic system.

MCT are rapidly emptied from the stomach, rapidly absorbed and metabolized and are a valuable source of energy during exercise. However ingestion of 30 g of MCT does not increase exercise performance.

EXERCISE ON PROTEIN REQUIREMENTS

Exercise, especially endurance exercise, results in increased oxidation of the branched-chain amino acids (BCAA) which are essential amino acids and cannot be synthesized in the body. Therefore increased oxidation would imply that the dietary protein requirement is increased. Dietary protein requirements for athletes involved in prolonged endurance training were higher than those for sedentary individuals.

Protein may contribute up to about 15% to energy expenditure in resting conditions. Although protein oxidation is increased during endurance exercise, the relative contribution of protein to energy expenditure remains small.

REVIEW QUESTIONS

1. Write a note on the muscle metabolic systems in exercise.

2. Discuss the nutrients used during muscle activity.

3. Write a note on the pre-event meal.

4. Discuss the effect of exercise on the requirement of carbohydrate and protein.

CRITICAL THINKING QUESTIONS

1. Why is a pre-game meal important?

2. Do athletes have to be careful about their diet?

16

NUTRIENT-DRUG INTERACTION

Did You Know?

- Drug absorption is excellent in proximal small intestine.

- Drugs like ampicillin, erythromycin, ferrous salts, isoniazid, etc. should be taken in a fasting state as absorption is increased.

- The chemical stability of a drug is affected by pH, enzyme activity, etc.

- Tetracycline prevents absorption of calcium, magnesium, iron or zinc.

- Most drugs must undergo biotransformation for excretion.

This chapter discusses the interactions among diet, nutritional status and drugs, the effect of drugs on food intake, body weight nutrient requirements and growth.

Medical conditions, age and gender dictate the drugs a patient may be taking and may influence a person's diet and nutritional status.

The term pharmacokinetics refers to the characteristics of a drug's absorption, distribution biotransformation and excretion; the term pharmacodynamics refers to the mechanism of drug action and the relationship between a drug's concentration at the active site and its pharmacological effects.

BIOAVAILABILITY

The term bioavailability refers to the extent to which a drug reaches its site of pharmacologic action. This definition includes the extent to which the drug reaches a fluid (e.g. blood or spinal fluid) that bathes the site of action and via which the drug can readily reach the site of action. The bioavailability of a drug depends directly on the extent to which the drug is absorbed and distributed to the site of action and depends inversely on the extent to which it is metabolized and excreted prior to arriving at the site of action.

ABSORPTION

The term absorption refers to the rate at which, and the extent to which, a drug leaves its site of administration. Factors that affect absorption include the route and site of drug administration; the site and area of absorption; the concentration of the drug at that site, the physical form of that drug, and the vehicle in which the drug is administered, local conditions that affect the drug's solubility and therefore its ability to reach a site of entry into the bloodstream or cerebrospinal fluid; and finally, the circulation at the site of absorption. Drug absorption is influenced by physico-chemical processes that influence both the extent and the rate at which the drug crosses the mucosal barrier and enters the bloodstream (Table 16.1).

Table 16.1   Sites of drug absorption and their characteristics

Site	Absorption	Characteristics
Mouth	Poor	Short residence time drug can make contact only if in solution.
Stomach	Poor	Small surface area, thick, mucus-covered, electrically resistant mucosa
Proximal small intestine	Excellent	Large surface area, thin, low-electrical-resistance mucosa can absorb both unionized and ionized drugs.

Food components affect drug absorption and bioavailability through three general mechanisms: physico-chemical interactions between drug and food components in the gut lumen. Alterations in gastric emptying time, and competitive inhibition between the drug and food components for absorption.

PHYSICO-CHEMICAL INTERACTIONS BETWEEN DIET INGREDIENTS AND DRUG INGREDIENTS

Physico-chemical interactions between diet ingredients and drug ingredients include absorption, complex formation, precipitation, and the effects of one interactant on another whereby stability of one or both is altered. Physico-chemical interactions require the simultaneous presence of the drug and food component at the site of interaction. Therefore, physico-chemical interactions occur in the gut lumen of orally and enterally administered drugs and in the reservoir or tubing containing the nutrient formula when drugs and the formula are administered simultaneously via parenteral or enteral routes. Hence drug and nutrient absorption is reduced.

Certain drugs chelate minerals and render both the drug and the mineral unavailable for absorption, e.g. doxycycline chelates with calcium, magnesium, iron or zinc. Other drugs like ciprofloxacin, penicillamine and thyroxine form stable complexes with iron.

GASTRIC EMPTYING AND DRUG ABSORPTION

A drug leaves the stomach at a rate that depends on gastric emptying. Gastric emptying, in turn, depends on the general functioning of the stomach and gastrointestinal tract (e.g. a person with autonomic neuropathy affecting the gut has very slow gastric emptying) on the physiological changes in the gastrointestinal tract in the fed versus the fasted stated, and on the effects of the drugs themselves on gastric emptying. In the fasting state or when little food is in the stomach, drugs usually leave the stomach rapidly and thus reach the small intestine.

Particles up to 1.1 mm in diameter are emptied from the stomach with food prior to the interdigestive phase, although there is great interindividual variability and not all particles of the same size may be released with food. Larger particles are released from the stomach during the interdigestive phases, with the very largest particles requiring "phase three contractions" of the stomach that occur well after a meal, when the stomach is empty. This phenomenon accounts for the observation that certain drug preparations are not cleared from the stomach until night, even though they may be consumed during the day if there is not sufficient time between meals and snacks to ensure complete emptying of the stomach. There is evidence that taking enteric coated pills or capsules with meals severely delays their delivery into the duodenum, sometimes as much as 10 to 12 hours. If given in multiple doses throughout the day, these drugs may accumulate in the stomach and be dumped into the duodenum at night when the stomach finally empties. This can be prevented by taking the drug on an empty stomach.

TRANSPORT ACROSS MEMBRANE

To be absorbed, drugs must be transported across the lipid membrane of the mucosal barrier, either by diffusion or by a specific carrier. Most drugs are absorbed in the small intestine and by simple diffusion that depends on the concentration gradient of the drug across the intestinal mucosa.

DIETARY EFFECTS ON DRUG FUNCTION

When a drug acts on processes that involve dietary components, that dietary component may affect that drug's functions. The most common ones are discussed.

Dietary Electrolytes and the Functions and Side Effects of Drugs Used to Treat Cardiovascular Disease

The digitalis glycosides block Na^+–K^+ ATPase. Hypokalaemia potentiates the ATPase block and leads to toxicity. Therefore, dietary maintenance of normal plasma potassium levels is critical for prevention of this side effect particularly if the patient is also taking potassium-losing diuretics.

A dose-dependent side effect of the thiazide diuretics which occurs in up to 3% of subjects, is development of glucose intolerance and overt diabetes mellitus maintenance of adequate plasma potassium and magnesium levels through diet or if necessary supplements may prevent development of this complication.

Dietary Effects on the Coumarin Anticoagulants

The coumarin anticoagulants (warfarin, coumadin, dicumerol) act through antagonism of vitamin K. Consumption of vitamin K-enriched foods or supplements may counteract the effects of the anticoagulants. Because the dosage of the coumarin antagonists is based on the patient's prothrombin time rather than on vitamin K-levels, it is more important to maintain a steady, consistent intake of vitamin K-containing foods while receiving anticoagulants, than it is to decrease intake.

Certain herbal leaves contain natural coumarins that can augment the effects of coumarins and should be avoided. Brussel sprouts and other cruciferous vegetables increase the catabolism of warfarin and thereby decrease its anticoagulant activities.

Interactions of Caffeine and Other Methyl Xanthines with Drug Function

Caffeine and other methyl xanthines have numerous effects on the function of other drugs, both through induction of p450 enzymes and through other pathways. For example, methyl xanthines enhance the cytotoxicity of alkylating agents by hastening passage through the G_2 phase of the cell cycle, thereby decreasing the time for repair of sequentially "Lit" DNA. *In vitro*, caffeine binds to DNA-intercalating drugs such as the antitumour drug doxorubicin, *in vivo* consequences of this binding are not known. Caffeine also decreases the effectiveness of a number of antiepileptic drugs, including phenobarbital and valproate, in preventing electroshock seizures in mice, probably through adenosine-mediated inhibition. These findings suggest that persons taking these drugs should probably not take caffeine-containing foods and beverages, or if they do, they should maintain their intake at a constant level.

Interactions of Diet with Hypolipidaemic Drugs

Although the hypolipidaemic effects of the hypolipidaemic drugs may be augmented by a lipid-lowering diet, the effects are usually additive and do not result from a specific interaction between diets and the drug. However fibrates bind to proteins known as peroxisome proliferator-activated receptors (PPARs). PPARs, members of the steroid hormone receptors, play a distinctive role in the regulation of fat and cholesterol metabolism.

DRUG EXCRETION

Drugs are primarily excreted via the kidneys and the gastro-intestinal tract. However, many drugs are excreted into breast milk, and non-electrolytes (e.g. ethanol) reach the same concentration in breast milk as they do in plasma.

Effects of Diet on Drug Excretion

Diet can affect drug excretion by altering either the clearance by the drug's organ of excretion or the drug's volume of distribution. The half-life of a drug in circulation is a direct function of the volume of distribution of the drug and an inverse function of its clearance.

The clearance of a drug depends on the amount of drug delivered to the organ of excretion and the extent to which the drug is extracted from the blood for excretion (the extraction ratio). In the case of renal excretion, clearance of the drug depends on the concentration of free drug in the plasma, on the glomerular filtration rate, and on the amount of drug secreted and/or reabsorbed in the tubules per unit rate.

LOW-PROTEIN DIETS AND HYPOALBUMINAEMIA

The concentration of free drug in the plasma depends on the chemical characteristics of the drug. For example, if a drug binds to albumin, a decrease in plasma albumin levels (due to inadequate protein intake or disease) may increase the proportion of free drug to bound drug. However, the effect on excretion of an increase in the unbound fraction of the drug may be counteracted by an increase in the drug's volume of distribution. An increase in distribution volume may occur for two reasons. First, hypoalbuminaemia leads to an increase in total body water.

Second, if the drug binds to particular tissues (e.g. adipose tissue), a larger proportion of the drug is sequestered by these tissues and therefore is unavailable for immediate excretion.

Thus, low protein diets can in theory affect renal clearance of drugs in potentially opposing ways. Acutely low protein meals decrease glomerular filtration rates, so less drug is available for excretion per unit time. However, tubular reabsorption may be more efficient, so the clearance of the drug may be decreased out of proportion to the decrease in filtration rate. Chronic low protein consumption with hypoalbuminaemia leads to an increase in free drug available for excretion, if the drug is bound to albumin and to alterations in the volume of distribution. However, the result of a low protein diet on actual clearance varies from drug to drug and depends on the severity and duration of the low-protein intake. For example, dietary protein restriction decreases clearance of the anti-gout drug allopurinol and promotes renal tubular reabsorption of its chief metabolite, oxypurinol.

Another effect of a low-protein diet, is to increase net renal excretion of base. The increased excretion of base is due to decreased production of acid residues (phosphate, sulphate and chloride) from protein metabolism, coupled with decreased renal secretion of hydrogen ions due to increased extracellular fluid volume. This effect lessens the rate of excretion of basic drugs such as the antibiotic gentamicin or the antiarrhythmic drugs procainamide and guanidine because less of the ionized form of these drugs is presented in the renal tubule and thus more of the drug is reabsorbed. A similar effect is produced by intake of antacids. In patients receiving basic drugs, urinary pH should be monitored, and if the pH is increasing, the drug dose should be reduced or an alternative drug considered.

DEHYDRATION AND STARVATION

Dehydration, and starvation without hypoproteinaemia, may significantly decrease the distribution volume of a drug because of decreases in the size of the aqueous compartment. Glomerular filtration rates may also be decreased. In addition, starvation decreases the distribution volume of lipophilic drugs because of a loss of adipose tissue. By contrast, people who are lean but not starved have a higher proportion of body water per kilogram than nonlean people, so the half-life of drugs that are primarily distributed in the aqueous compartment is prolonged in lean persons.

Competition between drugs or between drugs and nutrients for a common renal pathway can also change the rate of drug excretion. Renal tubular reabsorption of lithium chloride is increased in sodium depletion. Because of the potential for toxicity, a patient receiving lithium chloride should not be placed on a low-sodium diet unless close attention is paid to blood lithium levels.

EFFECT OF DRUGS ON FOOD INTAKE, BODY WEIGHT, NUTRIENT REQUIREMENTS, AND GROWTH

Alteration in Food Intake

Drugs can alter food intake by causing either a perversion or loss of appetite or an increase in appetite.

Drugs may suppress appetite either as their primary effect (obesity drugs) or as an unwanted side effect.

Virtually every drug can cause nausea and vomiting, either centrally through activation of the chemoreceptor trigger zone in the area postrema of the brain or, more commonly, peripherally through irritation of the gastrointestinal tract. Any drug that induces nausea is likely to reduce food intake and hence contribute to weight loss. Food aversion develops when eating the food is associated with development of nausea and vomiting. Food aversions develop in more than half of all patients who undergo cancer chemotherapy and usually affect consumption of two to four foods. The aversions are usually transient.

Drugs may also cause a dry or sore mouth and make eating difficult or painful which leads to avoidance of food intake. Patients may develop sore mouths because of drugs that cause stomatitis (gum inflammation and hyperplasia), delay in wound repair and healing, gingival bleeding, microbial infections and oral candidiasis. Dry mouth may be caused by drugs that decrease salivary flow.

Drugs can alter the taste of saliva or food by excretion in the saliva, often leading to a metallic taste; alteration or decrease in taste receptor function and alteration of olfactory function. Some drugs may not taste good or may have a disagreeable texture. Some drugs prevent the replication of taste bud cells.

Many drugs cause gastric irritation which results in disagreeable sensations in the stomach that leads to a loss of appetite, e.g. aspirin.

Some of the drugs, e.g. erythromycin, increase the rate of gastric emptying and the overall motility of the GI tract. Opioids cause considerable delays in gastric emptying and marked constipation so increasing dietary fibre becomes necessary.

Bezoars

Bezoars are, balls of unabsorbed material residing in the stomach formed from undissolved portions of drugs, masses of food matter, in particular plant fibre (phytobezoars). Bezoars tend to form in patients with achlorhydria, in whom tablets may not dissolve, and in those with poor gastric emptying often because of surgical intervention, in patients with autonomic neuropathy due to diabetes mellitus or in patients with alcoholism. Bezoars may cause feelings of gastric fullness, pain, nausea, vomiting and gastric outlet obstruction.

Drug-induced Maldigestion and Malabsorption

Drug-induced maldigestion and malabsorption may be due to interactions between the drug and the nutrient or to the changes in the gastrointestinal tract. For example cancer chemotherapeutic agents cause damage of the internal mucosa because of their antimitotic effect.

EFFECT OF DRUGS ON VITAMIN AND MINERAL STATUS REQUIREMENTS AND ACTIVITY

Drugs can affect vitamin and mineral status by interfering with absorption, metabolism and function. The antivitamin or mineral-related effects of drugs may be used intentionally in the treatment of disease or may be an unwanted side effect of drug therapy.

Antacids

By increasing stomach pH, antacids decrease the bioavailability of vitamin A, folate, thiamine and phosphate.

Vitamin A and Carotene Malabsorption

Absorption of vitamin A and carotenoids requires release of these compounds from dietary proteins via pepsin proteolysis. Pepsin is rapidly inactivated, and autoactivation of pepsin from pepsinogen is suppressed at pH 5. In addition, aluminium hydroxide absorbs pepsin above pH 3.

Thiamine degradation Thiamine is unstable at high pH and hence is inactivated when taken with antacids.

Folate malabsorption When the pH of the upper part of the jejunum is increased after ingestion of sodium bicarbonate, folate absorption is reduced.

Sodium overload Sodium overload with development of congestive heart failure can occur from intake of excessive sodium.

Phosphate depletion Dietary phosphate combines with aluminium and magnesium hydroxide to form insoluble aluminium and magnesium phosphates, which are excreted through the gastrointestinal tract. Hence, a phosphate depletion syndrome may develop, the problem is also compounded on a low phosphate diet. Effects of phosphate depletion include muscle weakness, paresthesia in the limbs, anorexia, haemolytic anaemia and convulsions. Myocardial depression may also occur.

Milk–alkali syndrome This occurs due to an excessive intake of alkali and milk.

Use of calcium salt supplements to delay age-related or medication (corticosteroid)-related bone loss or to prevent hypertension has the potential to cause the milk–alkali syndrome.

Magnesium overload Patients with chronic renal failure who are taking magnesium-containing antacids can develop magnesium intoxication. Symptoms of magnesium overload are nausea, vomiting, flushing, impaired respiratory function and partial or complete heart block.

Iron malabsorption Absorption of iron occurs when it is in the ferrous state, at low pH. The elevated pH induced by antacids may lead to iron aggregate formation and conversion of iron to its ferric form. In addition, aluminium hydroxide gels bind iron and thereby decrease its absorption.

H_2 blockers and proton pump inhibitors Histamine H_2 receptor antagonists and proton pump inhibitors can decrease stomach acid secretion. Hence, vitamin B_{12} is not released from dietary protein and becomes unavailable for absorption.

Omeprazole (given for ulcers) therapy decreases cyanocobalamin absorption.

Sulphasalazine is prescribed for inflammatory bowel disease. Sulphasalazine is a competitive inhibitor of intestinal folate transport.

Antituberculous agents

Vitamin D metabolism The drugs used in the treatment of tuberculosis, rifampin and isoniazid, affect vitamin D metabolism because of their effect on vitamin D hydroxylase.

Pyridoxine Isoniazid and other carbonyl compounds combine with pyridoxine to form hydrozones. These compounds are potent inhibitors of pyridoxal kinase which increases the possibility of peripheral neuritis, hence, pyridoxine supplementation is necessary during isoniazid therapy.

Niacin deficiency　Isoniazid can also cause a secondary niacin deficiency. This results due to an inhibition of the enzyme kynureninase which is necessary for the formation of nicotinamide nucleotide synthesis from tryptophan.

Anticonvulsants

Calcium and vitamin D metabolism　The anticonvulsant drugs phenytoin and phenobarbital can cause hypocalcaemia, rickets in children, and high turnover osteoporosis in adults. Phenytoin acts through three mechanisms. Decreased calcium absorption, possibly through inhibition of the synthesis of calcium-binding protein; stimulation of the catabolism of 25-hydroxy cholecalciferol, with a possible reduction in 24, 25-dihydroxy cholecalciferol production, and increased catabolism of vitamin K, with reduction in formation of the vitamin K-dependent proteins involved in calcium uptake by osteoblasts in particular osteocalcin. Levels of 1, 25 dihydroxycholecalciferol appear to be normal.

Folate metabolism　High doses of folic acid may counteract the anticonvulsant effects of phenobarbital, phenytoin and primidone. Patients who develop megaloblastic anaemia, while taking these drugs, may respond to folate supplementation with increased seizure frequency.

Vitamin K catabolism　Phenytoin and phenobarbital probably through their action on hepatic p450 enzymes, increase hepatic catabolism of vitamin K and decrease production of vitamin K-dependent proteins.

Bile acid sequestrants　Drugs that bind bile acids (e.g. cholestyramine and aluminium-containing antacids) may lead to malabsorption of fats and hence to malabsorption of fat-soluble vitamins, particularly vitamin A and E. Cholestryramine also binds folic acid.

Corticosteroids Corticosteroids administered chronically increase the need for vitamin B$_6$, calcium and vitamin D.

The following nutritionally significant consequences of long-term corticosteroid use are seen in almost all systems.

1. Gastrointestinal system

    Peptic ulceration

    Gastric or intestinal perforation

    Pancreatitis

2. Endocrine system and metabolism

    Hyperglycaemia

    Hyperlipidaemia

    Weight gain

    Altered distribution of fat tissue

    Hypertension

3. Skeletal system

    Osteoporosis

    Growth failure

    Fractures

4. Immune system

    Immunosuppression

Diuretics

Electrolyte imbalance Because diuretics act by altering renal excretion of sodium and potassium, they may lead to the development of electrolyte imbalance. Common drugs that cause potassium deficiency include thiazide, laxatives, etc. The potassium-losing diuretics deliver large amounts of sodium to the distal tubule, where part of the sodium is exchanged for potassium with loss of potassium. Drugs that cause potassium depletion may cause renal tubule damage.

Calcium status   Thiazide diuretics cause renal calcium retention and can cause hypercalcaemia. A greater bone mineral content is also seen. Hence, thiazide drugs have been suggested as a therapeutic drug in the management of osteoporosis.

Nondiuretic-antihypertensive agents   The antihypertensive drug diazoxide increases the proximal tubular reabsorption of sodium and can cause an overload of sodium.

Other non-nutritive components of food

Monosodium glutamate (MSG)   This is used as a flavour enhancer in many foods, especially chinese foods. MSG can cause mental retardation in infants and intestinal discomfort in adults.

Tartrazine   It is a colour additive used in both foods and drugs. Many are allergic to this dye.

Sulphites   These are added to foods, beverages and drugs as preservatives because of their antioxidant properties. They can cause severe allergic reactions.

Licorice   This causes sodium and water retention and increases potassium excretion.

Lactose   This is present in many drugs as an excipient. Lactose-intolerant patients have developed intestinal bloating and diarrhoea when taking these drugs in large quantities.

Aspartame   This is used as an artificial sweetener and may be detrimental to persons with phenylketonuria.

1. Discuss the gastric emptying and drug absorption process.

2. Write in detail about the dietary effects on drug function.

3. Discuss the process of drug excretion.

4. Write short notes on:

     i. Bioavailability

    ii. Absorption

   iii. Hypoalbuminaemia

    iv. Bezoars

     v. Phosphate depletion

    vi. Antituberculosis agents

   vii. Anticonvulsants

1. Why do drugs affect growth?

APPENDIX I

Table A1.1 Hormone-like compounds that help regulate the digestive process

Compound	Main location	Proposed main actions
Motilin	Upper small intestine	Stimulates intestinal motility
Pancreatic polypeptide	Pancreas	Inhibits pancreatic secretion, relaxes gall bladder
Somatostatin	Widespread	Inhibits glandular secretions and nerve transmissions in intestinal muscles
Neurotensin	Ileum	Inhibits stomach acid release and stomach emptying; stimulates pancreatic bicarbonate release
Enteroglucagon	Ileum, colon	Slows intestinal transit; stimulates intestinal mucosa
Vasoactive intestinal peptide (VIP)	Widespread	Dilates blood vessels; relaxes GI muscles; stimulates pancreatic, gall bladder, and intestinal secretion; inhibits stomach acid secretion
Substance P	Widespread	Dilates blood vessels, contracts GI muscles, stimulates pancreatic and salivary secretions
Enkephalins	Widespread	Inhibits nerve transmission in GI muscles

Table A1.2　Summary of digestive enzymes

Secretion origin	Enzyme	Substrate	Major end products
Salivary glands	Salivary amylase	Starch, glycogen	Maltose
Lingual glands	Lingual lipase	Short-chain triglycerides, medium-chain triglycerides	Fatty acids, monoglycerides
Stomach glands	Pepsin	Protein	Peptides, peptones
	Gastric lipase	Short-chain triglycerides, medium-chain triglycerides	Fatty acids, monoglycerides
Pancreas	Trypsin	Protein, peptides	Polypeptides, smaller peptides
	Chymotrypsin	Protein, peptides	Same as trypsin, more coagulating power for milk
	Carboxypeptidase	Polypeptides	Smaller peptides, free amino acids
	Pancreatic amylase	Starch, glycogen	Maltose
	Lipase	Triglycerides	Monoglycerides, free fatty acids
Intestinal wall	Aminopeptidase	Peptides	Amino acids, smaller peptides
	Maltase	Maltose	Glucose
	Sucrase	Sucrose	Glucose, fructose
	Lactase	Lactose	Glucose, galactose
	Enterokinase	Trypsinogen	Trypsin

Table A1.3   Digestion in practical terms

APPENDIX II

Table A2.1   Dietary fibre values for fibre-containing foods

Food group	Serving	Kilo calories	Grams of dietary fibre
Breads and cereals			
100% Bran	½ cup	75	84
Oatmeal (cooked)	1 cup	144	2.2
Whole-wheat bread	1 slice	60	1.4
Legumes, cooked			
Kidney beans	½ cup	110	7.3
Lima beans	½ cup	130	4.5
Vegetables, cooked			
Beans, green	½ cup	15	1.6
Broccoli	½ cup	20	2.2
Brussels sprouts	½ cup	30	2.3
Cabbage, red and white	½ cup	15	1.4
Carrots	½ cup	25	2.3

(Contd.)

Table A2.1 (Continued)

Food group	Serving	Kilo calories	Grams of dietary fibre
Cauliflower	½ cup	15	1.1
Corn	½ cup	70	2.9
Green pepper	½ cup	12	0.8
Green peas	½ cup	55	3.6
Kale	½ cup	20	1.4
Lettuce	1 cup	7	0.8
Parsnip	½ cup	50	2.7
Potato, with skin	1 medium	95	2.5
Tomato, chopped	½ cup	17	1.5
Fruits			
Apple	1 medium	80	3.5
Apricot, fresh	3 medium	50	1.8
Apricot, dried	5 halves	40	1.4
Banana	1 medium	105	2.4
Cantaloupe	¼ melon	50	1.0
Cherries	10	50	1.2
Dates, dried	3	70	1.9
Grapefruit	½ cup	40	1.6
Orange	1 medium	60	2.6
Peach	1 medium	35	1.9
Pineapple	½ cup	40	1.1
Prunes, dried	3	60	3.0
Raisins	¼ cup	110	3.1
Strawberries	1 cup	45	3.0

APPENDIX III

NUTRIENT CONVERSION FORMULAE FOR VITAMINS A AND E

Table A3.1 Formulae for converting forms of vitamin A into retinol equivalents (REs)

1. If retinol and β-carotene are given in μg

$$\mu g \text{ retinol} + \frac{\mu g \, \beta\text{-carotene}}{6} = RE$$

2. If retinol and β-carotene are given in IUs,

$$\frac{IU \text{ retinol}}{3.3} + \frac{IU \, \beta\text{-carotene}}{10} = RE$$

3. If β-carotene and carotenoids are given in μg,

$$\frac{\mu g \, \beta\text{-carotene}}{6} + \frac{\mu g \, \text{carotenoids}}{12} = RE$$

4. If retinol, β-carotene and carotenoids are given in μg,

$$\mu g \text{ retinol} + \frac{\mu g \, \beta\text{-carotene}}{6} + \frac{\mu g \, \text{carotenoids}}{12} = RE$$

Table A3.2 Formula for converting forms of vitamin E into α-tocopherol equivalents

α-tocopherol equivalents =

$$mg \, \alpha\text{-tocopherol} + \frac{mg \, \beta\text{-tocopherol}}{4}$$

$$+ \frac{mg \, \gamma\text{-tocopherol}}{10} + \frac{mg \, \alpha\text{-tocotrienol}}{3.3}$$

If only α-tocopherol is listed, multiply by 1.2 to determine α-tocopherol equivalents

Appendix IV

CLINICAL AND BIOCHEMICAL MEASURES OF NUTRITIONAL STATUS

Table A4.1 Clinical signs and symptoms of various nutrient deficiencies

Area of examination	Sign/symptoms	Potential nutrient deficiency
Hair	Alopecia	Zinc, essential fatty acids
	Easy pluckability	Protein, essential fatty acids
	Lackluster	Protein, zinc
	"Corkscrew" hair	Vitamin C, Vitamin A
	Decreased pigmentation	Protein, copper
Eyes	Xerosis of conjunctiva	Vitamin A
	Corneal vascularization	Riboflavin
	Keratomalacia	Vitamin A
	Bitot's spots	Vitamin A
GI tract	Nausea, vomiting	Pyridoxine
	Diarrhoea	Zinc, niacin
	Stomatitis	Pyridoxine, riboflavin, iron
	Cheilosis	Pyridoxine, iron
	Glossitis	Pyridoxine, zinc, niacin, folate, vitamin B_{12}
	Magenta tongue	Riboflavin
	Swollen, bleeding gums	Vitamin C
	Fissured tongue	Niacin
	Hepatomegaly	Protein

(Contd.)

Table A4.1   (Continued)

Area of examination	Sign/symptoms	Potential nutrient deficiency
Skin	Dry and scaling	Vitamin A, essential fatty acids, zinc
	Petechiae/ecchymoses	Vitamin C, vitamin K
	Follicular hyperkeratosis	Vitamin A, essential fatty acids
	Nasolabial seborrhoea	Niacin, pyridoxine, riboflavin
	Bilateral dermatitis	Niacin, zinc
Extremities	Subcutaneous fat loss	Kilocalories
	Muscle wastage	Kilocalories, protein
	Oedema	Protein
	Osteomalacia, bone pain, rickets	Vitamin D
	Arthralgia	Vitamin C
Haematologic	Anaemia	Vitamin B_{12}, iron, folate, copper, vitamin E, vitamin K
	Leukopenia, neutropenia	Copper
	Low prothrombin, prolonged clotting time	Vitamin K
Neurologic	Disorientation	Niacin, thiamine
	Confabulation	Thiamine
	Neuropathy	Thiamine, pyridoxine, chromium
	Parasthesia	Thiamine, pyridoxine, vitamin B_{12}
Cardiovascular	Congestive heart failure, cardiomegaly, tachycardia	Thiamine
	Cardiomyopathy	Selenium

Table A4.2 Biochemical indicators of good nutrition status

Nutrient or measurement	Test		Normal or acceptable levels	
			Men	Women
Iron	Haemoglobin (g/100 ml)		≥140.0	≥12.0
		Infants (under 2 years)	≥10.0	≥10.0
		Children (6–12 years)	≥11.5	≥11.5
		Pregnancy (2nd trimester)		≥11.0
		(3rd trimester)		≥10.5
Protein	Serum albumin (g/100 ml)		≥3.5	≥3.5
Normal lipid metabolism	Serum cholesterol (mg/100 ml)		<200	<200
	Serum triglyceride (mg/100 ml)		<250	<250
Normal carbohydrate metabolism	Serum glucose (mg/100 ml)		75–110	75–110
Sodium	Serum sodium (mEq/l)		130–150	130–155
Potassium	Serum potassium (mEq/l)		3.5–5.3	3.5–5.3

(Contd.)

Table A4.2 (Continued)

Nutrient or measurement	Test	Normal or acceptable levels	
		Men	Women
Vitamin A	Plasma vitamin A (μg/100 ml)	>20	>20
Vitamin C	Serum vitamin C (mg/100 ml)	$\geq$ 0.3	$\geq$ 0.3
Riboflavin	Erythrocyte glutathione peroxidase (% stimulation of activity by added riboflavin cofactor)	<20	<20
Vitamin B_6	Trytophan load test— increase in excretion of xanthurenic acid (mg/day)	<25	<25
Folate	Serum folate (ng/ml)	>6.0	>6.0
Thiamine	Urinary thiamine (μg/g creatinine)	>65	>65
Zinc	Plasma zinc (μg/100 ml)	80–115	80–115

APPENDIX V

Table A5.1   Intakes and signs of vitamin A needs and toxicity in adults

Retinol (in micrograms)	Retinol in IU$_a$*	Daily intake* category	Major signs
50	167	Deficiency	Night blindness, skin abnormalities, and xerophthalmia (dryness and thickening of the membranes of the eyelid and eye).
150	500	Marginal status	Cures night blindness and other visual abnormalities.
300	1000	Marginal status	Corrects most, if not all, effects of vitamin A deficiency in individuals.
600	2000	Satisfactory status	Provides adequate total body reserves in most persons in a population. Currently, the RDI of the World Health Organization.
1000	3333	Satisfactory status	Provides generous reserves for most persons in a population. Currently, the RDA for the adult male in the United States.

(Contd.)

Table A5.1   (Continued)

Retinol (in micrograms)	Retinol in IU$_a$*	Daily intake* category	Major signs
5400 to 150,000	18,000 to 500,000	Teratogenicity	Abortion; prominent facial and cranial abnormalities, including the brain, ears, eyes, jaw, mouth, and lips; and defects of the heart, kidney, and gastrointestinal tract.
7500 to 210,000	25,000 to 700,000	Chronic toxicity	Headache, loss of hair, cracking of lips, dry and itchy skin, enlarged liver, bone abnormalities, muscle and joint pain, and visual defects.
300,000 or more	1,000,000 or more	Acute toxicity	Nausea, vomiting, headache, dizziness, blurred vision, and muscular incoordination.

Table A5.2 Summary of the fat-soluble vitamins, their functions, deficiency conditions, and food sources

Vitamin	Major functions	Deficiency symptoms	Dietary sources	Toxicity symptoms
Vitamin A (retinol) and pro-vitamin A (carotenoids)	1. Adapt to dim light (formation of rhodopsin); colour vision 2. Promote growth 3. Prevent drying of skin and eye 4. Promote resistance to bacterial infection	1. Night blindness 2. Bitot's spots 3. Xerophthalmia 4. Poor growth 5. Dry skin (keratinization)	Vitamin A Liver Fortified milk Pro-vitamin A Sweet potatotes Spinach Greens Carrots Cantaloupe Apricots	Foetal malformations, hair loss, skin changes, pain in bones
D (chole- and ergocalciferols)	1. Facilitates absorption of calcium and phosphorus 2. Maintain optimum calcification of bone	1. Rickets 2. Osteomalacia	Vitamin D Fortified milk Fish oils	Growth retardation, kidney damage, calcium deposits in soft tissue
E (tocopherols)	1. Antioxidant: prevents oxidation of vitamin A and unsaturated fatty acids	1. Haemolysis of red blood cells 2. Nerve destruction	Vitamin E Vegetable oils Some greens Some fruits	Muscle weakness, headaches, fatigue, nausea, inhibition of vitamin K metabolism
K (phyllo- and menaquinones)	1. Form prothrombin and other factors for blood clotting	1. Haemorrhage	Vitamin K Green vegetables Liver	Anaemia and jaundice

Table A5.3   Summary of the water-soluble vitamins

Name and coenzyme	Major functions	Deficiency symptoms	People most at risk
Thiamine; TPP	TPP is involved with enzymes in glycolysis, the citric acid cycle, hexose monophosphate shunt; nerve function	Beriberi; nervous tingling, poor coordination, oedema, heart changes, weakness	Alcoholism, poverty
Riboflavin; FAD and FMN	Coenzymes involved in citric acid cycle, electron transport chain, fat breakdown; drug detoxifying pathways	Ariboflavinosis; inflammation of mouth and tongue, cracks at corners of the mouth, eye disorders	Possibly people on certain medications if no dairy products consumed
Niacin; NAD and NADP	Coenzymes involved in glycolysis, citric acid cycle, electron transport chain, fat synthesis, fat breakdown	Pellagra; diarrhoea, bilateral dermatitis, dementia	Severe poverty where corn is the dominant food; alcoholism
Pantothenic acid; coenzyme A, acylcarrier protein	Citric acid cycle, fat synthesis, fat breakdown	Using an antagonist causes tingling in hands, fatigue, headache, nausea	Alcoholism
Biotin, biocytin	Glucose production, fat synthesis, purine (part of DNA, RNA) synthesis	Dermatitis, tongue soreness, anaemia, depression	Alcoholism

(Contd.)

Table A5.3   (Continued)

Name and coenzyme	Major functions	Deficiency symptoms	People most at risk
Vitamin B_6, pyridoxine and other forms; PLP	Protein metabolism, neurotransmitter synthesis, haemoglobin synthesis, many other functions	Headache, anaemia, convulsions, nausea, vomiting, flaky skin, sore tongue	Adolescent and adult women; people on certain medications; alcholism
Folate (folic acid); THFA	DNA and RNA synthesis, amino acid synthesis, choline synthesis	Megaloblastic anaemia, inflammation of tongue, diarrhoea, poor growth, mental disorders	Alcoholism, pregnancy, use of certain medications
Vitamin B_{12} (cobalamins)	Folate metabolism, nerve function	Macrocytic anaemia, poor nerve function	Elderly due to poor absorption; vegans
Vitamin C (ascorbic acid)	Collagen synthesis, hormone synthesis, neurotransmitter synthesis	Scurvy; poor wound healing, pinpoint haemorrhages, bleeding gums, oedema	Alcoholism, elderly men living alone

Table A5.4   Summary of the major trace minerals

Mineral	Major functions	Deficiency symptoms	People most at risk	Dietary source	Results of toxicity
Iron	Part of haemoglobin, myoglobin and cytochromes; used for immune function.	Low serum ferritin levels; small, pale red blood cells; low blood haemoglobin and haematocrit values	Infants, preschool children, adolescents, women in child-bearing years, some endurance athletes.	Meats, spinach, seafood, broccoli, peas, bran-enriched breads	Toxicity is seen when children consume 200–400 mg in iron pills, and in people with haemochromatosis. In this case people overabsorb iron.
Zinc	Over 200 enzymes need zinc. These include enzymes involved in growth, immunity, alcohol metabolism, sexual development, and reproduction.	Skin rash, diarrhoea, decreased appetite and sense of taste, hair loss, poor growth and development, poor wound healing	Vegetarians, and women in general, the elderly	Seafoods, meats, greens, whole grains	Reduces iron and copper absorption; causes diarrhoea, cramps, and depressed immune functions.
Copper	Aids in iron metabolism; part of many enzymes, such as those involved in protein metabolism and hormone synthesis.	Anaemia, low white blood cell (neutrophil) count, poor growth.	Infants recovering from malnutrition, intestinal surgery patients, and overzealous supple-mentation of zinc.	Meats, liver, cocoa, beans, nuts, whole grains	Vomiting; nervous system disorders (Wilson's disease).

(Contd.)

Table A5.4   (Continued)

Mineral	Major functions	Deficiency symptoms	People most at risk	Dietary source	Results of toxicity
Selenium	Peroxide metabolism, as is part of glutathione peroxidase.	Muscle pain, muscle weakness, heart disease.	Unknown.	Meats, organ meats, eggs, fish, milk, seafoods; grains grown where selenium is high in the soil.	Nausea, vomiting, hair loss, weakness, liver disease.
Iodide	Part of thyroid hormone	Goitre; poor growth in infancy when mother is deficient in pregnancy.	None in north America, as salt is usually fortified.	Iodized salt, white bread, salt water fish, dairy products.	Inhibition of function of the thyroid gland.
Fluoride	increases resistance of tooth crystal to acidic erosion.	Increased risk of dental caries.	Areas where water is not fluoridated and dental treatments do not make up for this lack of fluoride.	Fluoridated water, toothpaste, dental treatments, tea, seaweed.	Stomach upset, mottling (staining) of teeth during development.
Chromium levels	May increase action of the hormone insulin.	High blood glucose after eating.	People on total parenteral nutrition, and perhaps elderly people with adult onset diabetes mellitus.	Vegetable oils, egg yolks, whole grains, pork; yeast.	Due to industrial contamination, not dietary excess.

(Contd.)

Table A5.4 (Continued)

Mineral	Major functions	Deficiency symptoms	People most at risk	Dietary source	Results of toxicity
Manganese	Part of some enzymes, such as those involved in carbohydrate metabolism.	None in humans.	Unknown.	Nuts, rice, oats, beans.	Unknown in humans.
Molybdenum	Part of enzymes, such as xanthine dehydrogenase.	None in humans.	Unknown	Beans, grains, and nuts.	Unknown in humans.

Table A5.5   A summary of water and the major minerals

Name	Major functions	Deficiency symptoms	People most at risk	Dietary sources	Results of toxicity
Water	Medium for chemical reactions, removal of waste products, perspiration to cool the body	Thirst, muscle weakness, poor endurance	Infants with a fever, elderly in nursing homes	As such and in foods	Probably only in mental disorders, headache, blurred vision, convulsions
Sodium	A major ion of the extracellular fluid, nerve transmission	Muscle cramps	People severely restricting sodium to lower blood pressure (250–500 mg/day)	Table salt, processed foods	High blood pressure in susceptible individuals
Potassium	A major ion of intracellular fluid, nerve transmission	Irregular heart beat, loss of appetite, muscle cramps	Use of potassium-wasting diuretics, poor diets seen in poverty and alcoholism	Vegetables, fruits, milk	Slowing of the heart beat, seen in kidney failure
Chloride	A major ion of the extracellular fluid, acid production in stomach	Convulsions in infants	No one, probably, if infant formula manufacturers control product quality adequately	Table salt, some vegetables	High blood pressure in susceptible people when combined with sodium

(Contd.)

Table A5.5   (Continued)

Name	Major functions	Deficiency symptoms	People most at risk	Dietary sources	Results of toxicity
Calcium	Bones, teeth, blood clotting, nerve transmission, muscle contractions, cell regulation	Poor intake probably increases the risk for osteoporosis	Women in general, especially those who constantly restrict their energy intake and consume few dairy products	Dairy products, canned fish, leafy vegetables, tofu, fortified orange juice	Very high intakes may cause kidney stones in susceptible people
Phosphorus	Bones, teeth, metabolic compounds such as ATP, ion of intracellular fluid	Probably none; poor bone maintenance possible	Elderly consuming very nutrient-poor diets, total vegetarians? Alcoholism?	Dairy products, processed foods, soft drinks	Induces high levels of parathyroid hormone in kidney failure; poor bone mineralization if calcium intakes are low
Magnesium	Bones, enzyme function, nerve and heart function	Weakness, muscle pain, poor heart function	People on thiazide diuretics, women in general	Wheat bran, green vegetables, nuts, chocolate	Causes weakness in kidney failure
Sulphur	Part of vitamins and amino acids, drug detoxification, acid–base balance	None	No one who meets their protein needs	Protein foods	None likely

APPENDIX VI

Tips for Avoiding Too Much Fat and Saturated Fat

1. Steam, boil, or bake vegetables. For a change, stirfry in a small amount of vegetable oil.

2. Season vegetables with herbs and spices rather than with sauces, butter, or margarine.

3. Try lemon juice on salad or use limited amounts of oil-based salad dressing.

4. Try whole-grain flours to enhance flavours when making baked goods with less fat and cholesterol-containing ingredients.

5. Replace whole milk with skim or low-fat milk in puddings, soups, and baked products.

6. Substitute plain low-fat yogurt, blender-whipped low-fat cottage cheese, or buttermilk in recipes that call for sour cream or mayonnaise.

7. Choose lean cuts of meat.

8. Trim fat from meat before and after cooking.

9. Roast, bake, or broil meat, poultry, or fish so fat drains away as the food cooks.

10. Remove skin from poultry before cooking.

11. Use a nonstick pan for cooking so added fat will be unnecessary; use a vegetable spray for frying.

12. Chill meat or poultry broth until the fat becomes solid. Spoon off the fat before using the broth.

13. Eat a vegetarian main dish at least once a week. Include fish (cooked without much added fat) in the diet every week.

14. Limit high-fat cheese intake.

Table A6.1   Reduction of high serum LDL cholesterol level

Reduce saturated fat intake	This is the best method and should be the major focus
Reduce cholesterol intake	This helps some people and is not harmful to anyone
Perform regular exercise	This may protect serum HDL cholesterol levels
Lose weight to attain a desirable body weight	This helps reduce serum triglyceride levels (if elevated)
Increase intake of soluble fibre	This binds cholesterol and bile acids in the small intestine to encourage their elimination via the colon rather than absorption into the bloodstream; there also may be other causes for the drop
Reduce total fat intake	This may help achieve the other goals

Risk Factors and Dietary Advice for Adults to Control Cholesterol

Goal: Total blood cholesterol <200 mg/dl*

LDL cholesterol <160 mg/dl

or <130 mg/dl with 2 or more risk factors

Risk factors:

If total blood cholesterol is >200 mg/dl and an individual has two or more of the following risk factors:

✿ Family history of coronary heart disease

✿ Smokes cigarettes

✿ Diabetes

✿ Obesity

*mg/dl represents milligrams per 100 millilitres Lof serum.

✿ Hypertension

✿ Low HDL cholesterol

✿ Male

Test for LDL cholesterol

Advice: If LDL cholesterol > 130 mg/dl:

✿ Reduce saturated fat intake to 10% of total kilocalories

✿ Reduce total fat intake to 30% of total kilocalories

✿ Reduce cholesterol intake to 300 mg/day

For 6 months

If unsuccessful (that is, LDL cholesterol > 130 mg/dl)

✿ Reduce saturated fat intake to 7% of total kilocalories

✿ Reduce cholesterol intake to 200 mg/day

For 6 months

Appendix VII

Table A7.1   Potential drug–nutrient interactions for some commonly used drugs

Drug	Use	Nutrient	Potential side effect
Alcohol	-	Thiamine, vitamin B_6, folate, and zinc	Poor absorption/poor utilization
Antacids	Reduce stomach acidity	Calcium, vitamin B_{12} and iron	Decreased absorption due to altered gastrointestinal pH
Anticogulants	Preventation of blood clots	Vitamin K	Poor utilization
Antihistamines	Treatment of allergies and nausea; as local anaesthetic	–	Weight gain
Beta-blocker (propanolol)	Decrease hypertension	(Cholesterol)	Some can increase serum cholesterol levels
Aspirin	Anti-inflammatory, pain reduction	Iron	Anaemia from blood loss
Cathartics (laxatives)	To induce bowel movements	Calcium, potassium	Poor absorption
Cholestyramine	Reducing blood cholesterol	Vitamins, A, D, E, K	Poor absorption
Cimetidine	Treatment of ulcers	Vitamin B_{12}	Poor absorption
Colchicine	Treatment of gout	Vitamin B_{12}, carotenes, and magnesium	Decreased absorption due to damaged intestinal mucosa

(Contd.)

Table A7.1   (Continued)

Drug	Use	Nutrient	Potential side effect
Corticosteroids	Anti-inflammatory	Zinc	Poor absorption
Furosemide	Potassium-wasting diuretic	Potassium and sodium	Increased loss
Isoniazid	Tuberculosis	Vitamin B_6	Poor utilization
Neomycin	Antibiotic	Fat, protein, sodium, potassium, calcium, iron, and vitamin B_{12}	Decreases pancreatic lipase, binds bile salts, and so interferes with absorption
MAO inhibitors	Antidepressant	Tyramine in aged foods	Hypertension caused by poor tyramine metabolism
Phenobarbital	Sedative; treatment of epilepsy	Vitamin D and folate	Reduced metabolism and utilization
Phenytoin	Treatment of epilepsy	Vitamin D and folate	Reduced metabolism and utilization
Tricyclic antidepressants	Antidepressant	–	Weight gain due to appetite stimulation

GLOSSARY

Absorptive cells A class of cells that line the villi finger projections in the small intestine and participate in nutrient absorption.

Achlorhydria A state of reduced acid production by the stomach, primarily resulting from loss of the acid-producing cells in the stomach associated with ageing.

Acid pH A pH less than 7, e.g. lemon juice.

Active absorption Absorption in which a carrier is used and ATP energy is expended. In this way, the absorptive cell can absorb nutrients, such as glucose, against a concentration gradient.

Adenosine triphosphate (ATP) The main energy currency for cells. ATP energy is used to promote ion pumping, enzyme activity, and muscular contraction.

Adipose (fat) cells Fat storing cells.

Adipsin A protein that appears to be made by adipose cells and acts as a communication link between adipose cells and the brain.

Aerobic Requiring oxygen.

Aldosterone A powerful hormone produced by the adrenal glands that acts on the kidney to cause sodium reabosorption and, in turn, water conservation.

Alimentary canal Gastrointestinal tract.

Alkaline (basic) pH A pH greater than 7. Baking soda in water yields an alkaline pH.

Allergy An immune response when antibodies react with a foreign substance (antigen).

Alpha bond A type of carbohydrate bond that can be digested by human intestinal enzymes.

Alpha-linolenic bond A fatty acid with 18 carbon atoms and 3 double bonds (omega-3).

Alveoli Small air sacs of the lung.

Amino acid The building block for proteins containing a central carbon atom with a nitrogen atom and other atoms attached.

Amylase A branched-chain polysaccharide made of glucose units.

Amylose A straight-chain digestible polysaccharide made of glucose units.

Anabolism The process of building compounds.

Anaerobic Not requiring oxygen.

Anthropometry The measurement of weight, lengths, circumferences, and thicknesses of the body.

Antibody Blood proteins that inactivate foreign proteins found in the body. This helps prevent infections.

Antidiuretic hormone (ADH) A hormone, secreted by the pituitary gland, that acts on the kidney to cause a decrease in water excretion.

Antioxidant A compound that can donate electrons to electron-seeking compounds.

Apoferritin A protein in the intestinal cell that binds with the ferric form of iron (Fe^{3+}) to form ferritin.

Apolipoproteins Proteins embedded in the outer shell of lipoproteins.

Arachidonic acid A fatty acid with 20 carbon atoms and four double bonds (omega-6)

Ariboflavinosis A condition resulting from a lack of riboflavin. The "a" stands for "without", and the "osis" stands for "a condition of".

Atherosclerosis A build-up of fatty material in the arteries, including those surrounding the heart.

Atom Smallest combining unit of an element.

Autoimmune Immune reactions against normal body cells; self against self.

Avidin A protein found in raw egg whites that can bind biotin and inhibit its absorption. Avidin is destroyed by cooking.

Basal metabolism The minimum energy the body requires to support itself when resting and when awake. To have basal metabolic rate (BMR) measured, a person must not have eaten in the previous 12 hours and be maintained in a warm, quiet environment during the measurement.

Beriberi The thiamine deficiency disorder characterized by muscle weakness, loss of appetite, nerve degeneration, and sometimes oedema.

Beta bond A type of carbohydrate bond that is not digested by human intestinal enzymes when it is part of a long chain of monosaccharides.

Beta-oxidation The breakdown of a fatty acid into numerous acetyl-CoA molecules.

BHA and BHT (Butylated hydroxyanisole and butylated hydroxytoluene) Two common synthetic antioxidants added to foods.

Bile A substance made in the liver and stored in the gall bladder; it is released into the small intestine to aid fat absorption.

Bile acids Emulsifiers synthesized by the liver and released by the gall bladder during digestion to aid in fat digestion.

Bioavilability The degree to which the amount of an ingested nutrient actually gets absorbed and so is available to the body.

Biochemical lesion Nutritional deficiency symptoms observed in the blood or urine, such as low levels of nutrient by-products or low enzyme activities, indicating reduced body function.

Biological value of a protein The body's ability to retain protein absorbed from a food.

B-lymphocytes White blood cells synthesized by lymph tissues and responsible for antibody production.

Body mass index Weight (in kilograms) divided by height squared (in metres). A value of 25 or greater shows obesity-related health risks.

Bomb calorimeter An instrument used to determine the kilocalorie content of a food.

Calcitriol The active hormone form of vitamin D; it contains a derivative of cholesterol as part of its structure.

Calmodulin A cell protein that, once it binds calcium, can influence the activity of certain enzymes in the cell.

Carbohydrate A compound containing carbon, hydrogen, and oxygen atoms; most are known as sugars and starches.

Carbohydrate-loading A process in which a 600 g carbohydrate intake (or 70% of total energy, whichever is larger) is consumed for 7 days before an athletic event in an attempt to increase muscle glycogen stores.

Carnitine A compound used to shuttle fatty acids from the cytosol of the cell into mitochondria.

Carotenes Pigment substance in plants that can often form vitamin A. Beta-carotene is the most active form.

Casein Proteins in milk that form hard curds. These are difficult for infants to digest.

Catabolic Breaking down compounds.

Cell A minute structure, the living basis of plant and animal organization. In animals it is composed of a cell membrane. Cells contain both genetic material and systems of synthesizing energy-yielding compounds. Cells have the ability to both take up compounds from and excrete compounds into their surroundings.

Cellulose A straight-chain polysaccharide of glucose molecules that is undigestible because of the presence of beta carbohydrate bonds.

Celsius A centigrade measure of temperature:

$$\text{Fahrenheit (F)} - 32) \times \frac{5}{9} = {}^{\circ}\text{C}$$

$$\text{Centigrade (C)} \times \frac{9}{5}) + 32 = {}^{\circ}\text{F}$$

Ceruloplasmin A copper-containing protein component of plasma that changes iron to Fe^{3+} (ferric form) so it can bind with apoferritin.

Chemical score A ratio comparing the essential amino acid content of the protein in a food with the essential amino acid content in a reference protein, such as the one established by the Food and Agriculture Organization. The lowest ratio for an essential amino acid is the chemical score.

Cholesterol A waxy lipid; it has a structure containing multiple chemical rings.

Chronic Long-standing, developing over time; slow to develop or resolve. When referring to disease, this indicates that the disease progress, once developed, is slow and tends to remain; a good example is heart disease.

Chylomicrons Dietary fat surrounded by a shell of cholesterol, phospholipids, and protein. These are made in the intestine after fat absorption and travel through the lymphatic system to the bloodstream.

Chyme A mixture of stomach secretions and partially digested food.

***cis* isomer** An isomer form seen in compounds with double bonds, such as fat, where the hydrogen on both sides of the double bond lie on the same side of that bond.

Citric acid cycle A pathway that breaks down acetyl-CoA, yielding carbon dioxide, $FADH_2$, NADH, and GTP. The pathway can also be used to synthesize compounds.

Clinical lesion Nutritional deficiency sign seen on physical examination.

Complement A group of serum proteins involved in immune responses, such as phagocytosis and the destruction of bacteria.

Complete proteins Proteins that contain ample amounts of all nine essential amino acids.

Compound A group of different types of atoms bonded together in definite proportion (*See* molecule).

Connective tissue Tissue that holds different structures in the body together.

Cortisol A hormone made by the adrenal gland that, among other functions, stimulates the production of glucose from amino acids.

Crude fibre Remains of dietary fibre after acid and alkaline treatment. This consists of primarily cellulose and lignin.

Cytochromes Electron-accepting compounds that participate in the electron transport chain.

Deamination The removal of an amine group from an amino acid.

Dietary fibre Substances in food (essentially all from plant) that are not digested by the processes present in the stomach and small intestine.

Dietician *See* registered dietician.

Digestibility The proportion of food substances eaten that can be broken down in the intestinal tract and absorbed into the bloodstream.

Direct calorimetry A method to determine energy use by the body by measuring heat that emanates from the body, usually using an insulated chamber.

Disaccharides Class of sugars formed by the chemical bonding of two monosaccharides.

Duodenum The first 12 inches (30 centimetres) of the small intestine.

Eicosanoids Hormone-like compounds synthesized from polyunsaturated fatty acids. Within this class of compounds are prostaglandins, thromboxanes, and leukotrienes.

Eicosapentaenoic acid (EPA) An omega-3 fatty acid with 20 carbon atoms and 5 double bonds; present in fish oils.

Electron A part of an atom that is negatively charged. Electrons orbit the nucleus.

Electron transport chain A series of reactions using oxygen that converts NADH and $FADH_2$ into free NAD and FAD, yielding water and ATP.

Elements Substances that cannot be broken down further by using ordinary chemical procedures.

Enzyme A compound that speeds the rate of a chemical reaction but is not altered by the chemical reaction. Almost all enzymes are proteins.

Epithelial cells The surface cells that line the outside of the body and all external passages within it.

Eythropoietin A protein secreted by the kidneys that enhances red blood cell synthesis and stimulates red blood cell release from bone marrow.

Essential Having no obvious, external cause.

Essential amino acids Amino acids not efficiently synthesized by humans and that must therefore be included in the diet. There are nine essential amino acids.

Essential fatty acids Fatty acids that must be present in the diet to maintain health. These are linoleic acid and α-linolenic acid.

Estimated safe and adequate daily dietary intake (ESADDI) Nutrient intake recommendations first made in 1980 by the Food and Nutrition Board. A range for intake of some nutrients was given as not enough information was available to set a Recommended Dietary Allowance (RDA).

Extracellular fluid Fluid present outside the cells; this includes intravascular and interstitial fluids.

Extracellular space The space between cells.

Facilitated absorption Absorption where a carrier is used to shuttle substances into the absorptive cells, but no energy is expended. A concentration gradient higher in the intestinal lumen than in the absorptive cell drives the absorption.

Fasting hypoglycaemia Low blood sugar that follows a day or so of fasting.

Fatty acids Acids found in lipids, composed of carbon atoms bonded by hydrogen atoms, with an acid group

$$(-\overset{\displaystyle O}{\overset{\displaystyle \|}{C}}-OH)$$ at one end.

Ferritin A protein compound that serves as the storage form of iron in the blood and tissues.

Flavin adenine dinucleotide (FAD) A hydrogen carrier in the cell, synthesized from the vitamin riboflavin.

Fluoroapatite Tooth crystals containing fluoride ions that are relatively acid-resistant.

Free erythrocyte protoporphyrins (FEP) Immature forms of red blood cells released from the bone marrow. An increased serum level of FEP reflects a decreased ability to make red blood cells and suggests iron-deficiency anaemia. Lead poisoning also raises blood FEP levels.

Free radical Short-lived form of compounds that exist with an unpaired electron in their outer electron shell. This causes it to have an electron-seeking nature, which can be very destructive to electron-dense areas of a cell, such as DNA and cell membranes.

Fructose A monosaccharide with six carbons that form a five-membered ring with oxygen in the ring; found in fruits and honey.

Galactose A six-carbon monosaccharide; an isomer of glucose.

Galactosemia A disease characterized by the build-up of the monosaccharide, galactose, in the bloodstream resulting from the inability of the liver to metabolize it. If present at birth and left untreated, this causes severe growth and mental retardation in the infant.

Gastrointestinal (GI) tract The main sites in the body used for digestion and absorption of nutrients. It consists of the mouth, oesophagus, stomach, small intestine, large intestine, rectum, and anus.

Glucagon A hormone made by the pancreas that stimulates the breakdown of glycogen in the liver into glucose; this raises the blood glucose level. Glucagon also performs other functions.

Gluconeogenesis The production of new glucose molecules by metabolic pathways in the cell. The source of the carbon atoms for these new glucose molecules is usually amino acids.

Glucose A six-carbon atom carbohydrate found as such in blood, and in table sugar bound to fructose; also known as dextrose, it is one of the simple sugars.

Glycaemic index A ratio used to measure the relative ability of a carbohydrate to raise blood glucose levels as opposed to the ability of white bread (or glucose) to raise blood glucose levels.

Glycerol A carbohydrate containing three hydroxyl groups; used to help form triglycerides.

Glycogen A carbohydrate made of multiple units of glucose containing a highly branched structure; the storage form of carbohydrate for muscle and liver; sometimes known as animal starch.

Glycolysis The pathway that results in the breakdown of glucose into pyruvate (lactate) molecules.

Glycosidic bond The covalent bond formed between two monosaccharides when a water molecule is lost.

Goitre An enlargement of the thyroid gland often caused by a lack of iodide in the diet.

Goitrogens Substances in food that interfere with thyroid hormone metabolism and so may cause goitre if consumed in large amounts.

Gram Measure of weight in the metric system. One gram equals 1/28 of an ounce.

Gum A dietary fibre containing chains of galactose, glucuronic acid, and other monosaccharides; characteristically found in exudates from plant stems.

Harris–Benedict equation An equation that predicts resting metabolic rate based on a person's weight, height, and age.

Haematocrit The percentage of total blood volume occupied by red blood cells.

Haem iron Iron provided from animal tissues as products of haemoglobin and myoglobin. Approximately 40% of the iron in meat is haem iron. This is readily absorbed.

Hemicellulose A dietary fibre containing xylose, galactose, glucose, and other monosaccharides bonded together.

Haemosiderin An insoluble iron-protein compound found in the liver. Haemosiderin level increases as the amount of iron in the liver exceeds the storage capacity of ferritin.

Haemochromatosis A disorder of iron metabolism characterized by increased iron absorption, increased saturation of iron-binding proteins, and deposition of haemosiderin in the liver tissue.

Haemoglobin The iron-containing protein in the red blood cell that carries oxygen to the cells and carbon dioxide away from the cells. It is also responsible for the red colour of blood.

Haemolysis A breakdown of red blood cells caused by the destruction of the red blood cell membranes.

Hexose A general term describing a carbohydrate containing six carbon atoms.

High density lipoprotein HDL) The lipoprotein synthesized by the liver and intestine that picks up cholesterol from dying cells and other sources and transfers it to the other lipoproteins in the bloodstream. A low HDL level increases the risk for heart disease.

Histamine A breakdown product of the amino acid histidine that stimulates acid secretion by the stomach and other effects on the body, such as contraction of smooth muscles, increased nasal secretions, relaxation of blood vessels, and changes in relaxation of airways.

Hormone A compound secreted into the bloodstream that acts to control the function of distant cells. Hormones can be either protein-like or fat-like, such as insulin or oestrogen.

Hydrogenation The addition of hydrogen atoms to the double bonds of polyunsaturated and mono-unsaturated fatty acids to reduce the extent of unsaturation. This process turns liquid vegetable oils into solid fats.

Hydrophilic Attracts water (literally, "water-loving").

Hydrophobic Repels water (literally, "water-fearing").

Hydroxyapatite A compound composed of calcium and phosphate that is deposited into the bone protein matrix to give bone strength and rigidity.

Hypercalcaemia A high level of calcium in the bloodstream. This can lead to loss of appetite, calcium deposits in organs, and other health problems.

Hypercarotenaemia High level of carotene in the bloodstream, usually caused by a diet high in carrots or yellow squash.

Hyperglycaemia High blood glucose levels, above 140 mg per 100 ml of blood.

Hypertension A condition in which blood pressure remains persistently elevated, especially when the heart is between beats.

Hypertrophy An increase in cell size.

Hypochromic Pale red blood cells lacking sufficient haemoglobin (often caused by an iron deficiency).

Hypoglycaemia Low blood glucose levels, below 40 to 50 mg per 100 ml of blood.

Hypothalamus A part of the brain that contains cells that play a role in the regulation of hunger, respiration, body temperature, and other body functions.

Ileum The last half of the small intestine.

Incomplete proteins Lack of an ample amount of one or more essential amino acids to support human protein needs.

Index of nutritional quality (INQ) The numerical value of nutrient density.

Indirect calorimetry A method to measure the energy output by the body by measuring oxygen uptake and/or carbon dioxide output. Formulae are then used to convert these gas exchange values into kilocalorie use.

Inorganic Free of carbon atoms bonded to hydrogen atoms.

Insoluble fibres Fibres that, for the most part, do not dissolve in water nor are digested by bacteria in the large intestine. These include cellulose, some hemicelluloses, and lignin.

Insulin A hormone produced by the beta cells of the pancreas. Insulin increases the synthesis of glycogen in the liver and the movement of glucose from the bloodstream into muscle and adipose cells, among other processes.

Intermediate density lipoprotein (IDL) The product formed after a very low density lipoprotein (VLDL) has most if its triglyceride removed.

International unit (IU) A crude measure of vitamin activity, often based on the growth rate of animals. Today these units have been replaced by more precise mg and µg quantities.

Interstitial fluid Fluid between cells.

Intracellular fluid Fluid contained within a cell.

Intrinsic factor A protein-like compound produced by the stomach that enhances vitamin B_{12} absorption.

Ionic bond A union between two atoms formed by an attraction of a positive ion to a negative ion, as seen in table salt (Na^+ Cl^-).

Isomer Different chemical structures for compounds that share the same chemical formula.

Kcal The heat needed to raise 1,000 grams (1 litre) of water through 1 degree Celsius.

Ketone bodies Products of acetyl-CoA (fat) metabolism containing three to four carbon atoms: acetoacetic acid, beta-hydroxybutyric acid, and acetone. These contain a ketone group, hence, the name.

Ketosis The condition of having high levels of ketones in the bloodstream.

Kilojoule A measure of work in which one kilojoule equals the work needed to move one kilogram a distance of one metre with the force of one Newton. 1 kilocalorie equals 4.8 kilojoules.

Kwashiorkor A disease, seen primarily in young children, in which sufficient kilocalories are consumed, but not a sufficient amount of protein. The child will suffer from oedema, poor growth, weakness and an increased susceptibility to infections.

Lactic acid A three-carbon acid formed during anaerobic cell metabolism; a partial breakdown of glucose.

Lecithin A phospholipid containing two fatty acids, a phosphate group, and a choline molecule.

Lean body mass The part of the human body which is free of all but essential body fat. About 2% of body fat is essential to retain. The rest of the fat in the body represents storage and so is not part of lean body mass. Lean body mass includes muscle, bone, organs, connective tissue, skin, and other body parts.

Lignin An insoluble fibre made up of a multiringed alcohol (non-carbohydrate) structure.

Limiting amino acid The essential amino acid in the lowest concentration in a food in comparison with the body's need.

Linoleic acid A fatty acid with 18 carbon atoms and two double bonds; omega-6.

Lipase Fat-digesting enzymes; linguinal lipase is produced by the tongue, gastric lipase by the stomach, and pancreatic lipase by the pancreas.

Lipid A compound containing carbon, hydrogen, oxygen, and, sometimes, other atoms. Lipids dissolve in ether or benzene and are known as fat and oils.

Lipofuscin (ceroid pigments) Signs of accumulation of lipid breakdown products, often seen as brown spots on the skin.

Lipogenesis The building of fatty acids using derivatives of acetyl-CoA molecules.

Lipogenic Creating lipid. The liver is the major lipogenic organ in the human body.

Lipolysis The breakdown of lipid.

Lipoprotein A compound found in the bloodstream containing a core of lipids with a shell of protein, phospholipid, and cholesterol.

Lipoprotein lipase An enzyme attached to the outside of the cells that line the bloodstream; it breaks down triglycerides into free fatty acids and glycerol.

Litre A measure of volume in the metric system. One litre equals 0.96 quarts.

Long-chain fatty acids Fatty acids that contain more than 12 carbon atoms.

Low density lipoprotein (LDL) The product of the intermediate density lipoprotein (IDL) containing primarily cholesterol. An elevated level is strongly linked to heart disease.

Lysosome A cellular body that contains digestive enzymes for use inside the cell.

Lysozyme A substance produced by a variety of cells in the body that can destroy bacteria.

Major mineral A mineral vital to health that is required in the diet in amounts greater than 100 mg per day.

Malnutrition Failing health that results from a long-standing dietary intake that fails to meet nutritional needs.

Maltose Glucose bonded to glucose; a simple sugar.

Mannitol An alcohol derivative of fructose.

Marasmus Results in a person consuming insufficient protein and kilocalories; usually seen in infancy. It is equivalent of protein energy malnutrition in adults. The person will have little or no fat stores and show muscle wasting and weakness. Death from infections is common.

Medium-chain fatty acids Fatty acids that contain 8–12 carbon atoms.

Menaquinones Forms of vitamin K that come from animal food sources or bacterial synthesis.

Metabolism Chemical reactions that occur in the body, enabling cells to release energy from foods, convert one substance into another, and prepare end products for excretion.

Metallothionein Protein that binds and regulates the release of zinc and copper (and other positive ions) in intestinal and liver cells.

Metre A measure of length in the metric system, one metre equals 39.4 inches.

Micelles An emulsification product in which individual emulsifiers organize with their hydrophobic parts to the centre of micelle and their hydrophilic parts to the outside. Lipids are attracted to the centre area and water is attracted to the outside periphery.

Minerals Elements used in the body to promote chemical reactions and help form body structures.

Molecule A group of like or unlike atoms chemically combined (*See* compound).

Monoglyceride A breakdown product of a triglyceride consisting of one fatty acid bonded to the carbohydrate glycerol.

Monosaccharide A single sugar, such as glucose, that is not broken down further during digestion.

Monounsaturated fatty acid A fatty acid containing one $C=C$ double bond.

Mottling Discoloration or marking of the surface of teeth caused by a high fluoride content.

Mucilage A dietary fibre consisting of chains of galactose, mannose, and other monosaccharides; characteristically found in seaweed.

Mucopolysaccharides Substances containing protein and carbohydrate parts; found in bone and other ogans.

Mucus A thick fluid secreted by glands throughout the body. It contains a compound that has both carbohydrate and protein parts (glycoprotein). It acts as a lubricant and means of protection for cells.

Myocardial infarction Death of part of the heart muscle.

Net protein utilization Biological value of a protein multiplied by digestibility of that protein.

Neutrophil The major form of white blood cells, comprising 55 to 65% of the total.

Nicotinamide adenine dinucleotide A hydrogen carrier that represents a potential form of energy; made from the vitamin niacin.

Non-essential amino acids Amino acids that can be readily made by the body. There are 11 non-essential amino acids found in foods.

Non-haem iron Iron provided from plant sources and animal tissues other than part of haemoglobin and myoglobin. Non-haem iron needs to be changed to Fe^{2+} before absorption; less efficiently absorbed than haem iron.

Nonpolar A compound with no charges present; no positive or negative poles present.

Nucleus The core of an atom; it consists of protons and neutrons.

Nutrient density The ratio formed by dividing a food's contribution to the needs for a nutrient by its contribution to kilocalorie needs. When the contribution to nutrient needs exceeds that of kilocalorie needs, the food is considered to have a favourable nutrient density.

Nutrients Chemical substances in food that nourish the body by providing energy, building materials, and factors to regulate needed chemical reactions in the body. The body either cannot make these nutrients or can't make them fast enough for its needs.

Nutritional status The nutritional health of a person as determined by anthropometric measures (height, weight, circumferences, and so on), biochemical measures of nutrients or their by-products in blood and urine, a clinical (physical) examination, and a dietary analysis (ABCD).

Oedema The build-up of fluid in extracellular spaces.

Oligosaccharides Carbohydrates containing three to ten monosaccharide units.

Omega-3 fatty acid A fatty acid with its first double bond starting at the third carbon atom from the methyl end ($—CH_3$).

Omega-6 fatty acid A fatty acid with its first double bonds starting at the sixth carbon atom from the methyl end ($—CH_3$).

Organ A group of tissues designed to perform a specific function, for example, the heart. It contains muscle tissue, nerve tissue, and so on.

Osmotic pressure The pressure needed to be exerted to keep particles in a solution from drawing liquid across a semipermeable membrane.

Osteomalacia Adult rickets. A vitamin D deficiency disease that causes weak bones and increases fracture risk.

Osteoporosis A bone disease that developes primarily after menopause in women and is characterized by a decrease in bone density.

Overnutrition A state in which nutritional intake exceeds the body's needs.

Oxidize To lose an electron or gain an oxygen atom.

Oxidizing agent A compound capable of capturing an electron from another compound or supplying an oxygen atom to a compound.

Palatable Pleasing to taste.

Passive absorption Absorption that uses no energy. It requires permeability for the substance through the wall of the small intestine and a concentration gradient higher in the lumen of the small intestine than in the absorptive cell.

Pectin A dietary fibre containing chains of galacturonic acid and other monosaccharides; characteristically found between plant cell walls.

Pellagra A disease characterized by inflammation of the skin, diarrhoea, and eventual mental incapacity resulting from the lack of the vitamin niacin in the diet.

Pepsin A protein-digesting enzyme produced by the stomach.

Peptide bond A bond formed by the reaction of an amine group with an acid group while splitting off a water molecule. This is the main bond that links amino acids in a protein.

Peptides A few amino acids bonded together; often two to four.

Peptones A partial breakdown product of proteins.

Percentile Classification of a measurement of a group into divisions of 100.

Peristalsis A coordinated muscular contraction that is used to propel food down the gastrointestinal tract.

Pernicious anaemia The anaemia that results from a lack of vitamin B_{12} absorption. It is pernicious (deadly) because of the associated nerve degeneration that can result in eventual paralysis and death.

pH A measure of the hydrogen ion concentration in a solution.

Phagocytosis/pinocytosis A form of active absorption in which the absorptive cell forms an indentation, and particles or fluids entering the indentation are then engulfed by the cell.

Phylloquinone A form of vitamin K that comes from plants.

Physiological anaemia The normal increase in blood volume in pregnancy that ends up diluting the quantity of red blood cells, resulting in anaemia; also called haemo-dilution.

Phytic acid A constituent of plant fibres that binds positive ions to its multiple phosphate groups.

Phytobezoars A pellet of fibre characteristically found in the stomach.

Plaque A cholesterol-rich substance deposited in the blood vessels. It also contains various white blood cells, other lipids, and eventually calcium.

Polar A compound with distinct positive and negative charges (poles) on it. These charges act like poles on a magnet.

Polypeptide Fifty to 100 amino acids bonded together.

Polysaccharides Carbohydrates containing many glucose units, up to 3000 or more.

Polyunsaturated fatty acid A fatty acid containing two or more $C=C$ double bonds.

Pool The amount of a mineral stored within the body that can be readily mobilized when needed.

Positive balance A state in which a nutrient intake exceeds losses. This causes a net gain of the nutrient in the body, such as when tissue protein is gained during growth. The opposite of this would be negative balance, where losses exceed intake, as in cases of starvation.

Primary structure of a protein The order of amino acids in the protein molecule.

Protein Compounds made of amino acids, containing carbon, hydrogen, oxygen, nitrogen, and sometimes sulphur atoms, in a specific configuration.

Protein efficiency ratio A measure of protein quality determined by the ability of a protein to support the growth of a young rat.

Protein-energy malnutrition (PEM) This results when a person regularly consumes insufficient amounts of kilocalories and protein. The deficiency eventually results in body wasting and an increased susceptibility to infections.

Raffinose An indigestible oligosaccharide containing three monosaccharide units.

Rancid Containing products of decomposed fatty acids; these yield off-flavours and odours.

Recommended Dietary Allowances (RDA) Recommended intake for nutrients that meet the needs of almost all people of similar age and gender. These are established by the Food and Nutrition Board of the National Academy of Sciences.

Requirement The amount of a nutrient required by one person to maintain health. This varies between individuals. We do not know our individual requirements for each nutrient.

Resting metabolic rate Essentially the same as the basal metabolic rate, but the subject does not need to meet the strict conditions used for a basal metabolic rate determination. Today, both terms are often used interchangeably.

Retinoids Forms of preformed vitamin A; one source is animal foods like liver.

Reverse transport of cholesterol The process by which cholesterol is picked up by HDL molecules and transferred to other lipoproteins that can dispose of it in the liver.

Rhodopsin A protein involved in vision; it is made in the eye and incorporates a protein called opsin and a form of vitamin A; especially important in night vision.

Ribose A five-carbon sugar found in genetic material, specifically RNA.

Rickets A disease characterized by softening of the bones because of poor calcium deposition. This deficiency disease arises from lack of vitamin D activity in the body.

Risk factor A characteristic or a behaviour that contributes to the chances of developing an illness.

Satiety A state in which there is no longer a desire to eat.

Saturated fatty acid A fatty acid containing no C=C double bonds.

Scavenger pathway for cholesterol uptake A process by which LDL molecules (cholesterol-containing) are taken up by scavenger cells imbedded in the blood vessels.

Scurvy The deficiency disease that results after a few weeks of consuming a diet free of vitamin C; pinpoint haemorrhages are an early clinical sign.

Secondary deficiency A deficiency caused not by lack of the vitamin in question, but by lack of a substance that is needed for that vitamin to function.

Secondary structure of a protein The interactions (bonds) formed between amino acids placed close together in the primary structure.

Semi-essential amino acids Amino acids that, when consumed, spare the need to use an essential amino acid for their synthesis. Tyrosine in the diet, for example, spares the need to use phenylalanine for its synthesis.

Sequesterants Compounds that bind free metal ions. By doing so, they reduce the ability of ions to cause rancidity in compounds containing fat.

Serotonin A neurotransmitter synthesized from the amino acid tryptophan that appears to both decrease the desire to eat carbohydrates and induce sleep.

Set point Often refers to the close regulation of body weight. It is not known what cells control this set point nor how it actually functions in weight regulation. There is no doubt, however, that there are mechanisms that help regulate weight.

Short-chain fatty acids Fatty acids that contain fewer than eight carbon atoms.

Soluble fibres Fibres that either dissolve or swell when put into water or are metabolized (fermented) by bacteria in the large intestine. These include pectins, gums, mucilages, and some hemicelluloses.

Solvent   A substance that other substances dissolve in.

Sorbitol   An alcohol derivative of glucose.

Stachyose   An indigestible oligosaccharide with four monosaccharide units.

Starch   A carbohydrate made of multiple units of glucose attached together in a form the body can digest; also known as complex carbohydrate.

Sucrose   Fructose bonded to glucose; table sugar.

Superoxide dismutase   An enzyme that can neutralize a superoxide-free radical.

Symptom   A change in health status noted by the person with the problem, such as a stomach pain.

Tertiary structure of a protein   The three-dimensional structure of a protein, formed by interactions of amino acids placed far apart in the primary structure.

Thermic effect of food   The increase in kilocalorie use that occurs during the digestion, absorption, and metabolism of energy-yielding nutrients.

"Thrifty" metabolism   A metabolism that characteristically uses less kilocalories than normal, such that the risk of weight gain and obesity is enhanced.

Thromboxane A_2   A stimulus for blood clotting made by particles (platelets) in the bloodstream.

Thyroid-stimulating hormone   A hormone that regulates the uptake of iodide by the thyroid gland and is secreted in response to low levels of circulating thyroid hormone.

Tissue   A group of cells designed to perform a specific function; nerve tissue is an example.

T-lymphocyte   White blood cells synthesized by the thymus gland and responsible for recognition of foreign substances (for example, bacterial cells) in the body.

Tocopherols   The chemical name for some forms of vitamin E.

Trabecular bone   The spongy, inner matrix of bone, found primarily in the spine, pelvis, and ends of bones.

Trace mineral   A mineral vital to health that is required in the diet in amounts less than 100 mg per day.

Transamination   The transfer of an amine group from an amino acid to a carbon skeleton to form a new amino acid.

Triglyceride   The major form of lipid in food. It is composed of three fatty acids bonded to the carbohydrate glycerol.

Trypsin   A protein-digesting enzyme secreted by the pancreas (in a zymogen form) that acts in the small intestine.

Very low density lipoprotein (VLDL)   The lipoprotein that initially leaves the liver. It carries both the cholesterol and lipid newly synthesized by the liver.

Villi Finger-like protrusions into the small intestine that participate in digestion and absorption of foodstuff.

Vitamins Compounds needed in very small amounts in the diet to help regulate and support chemical reactions in the body.

Xerophthalmia A cause of blindness that results from infection of the eye resulting from a vitamin A deficiency. The specific cause is a lack of mucous production by the eye, which then leaves it more vulnerable to surface dirt and bacterial infections.

Xylitol An alcohol derivative of the five-carbon monosaccharide, xylose.

Zymogen An inactive form of an enzyme.

REFERENCES

Fox, Stuart, H. Ira, 2002. *Human Physiology,* 7th edn. McGraw-Hill., New Delhi.

Garrow, J.S., James, W.P.T. and Ralph, A. 2000. *Human Nutrition and Dietetics.* Churchill Livingstone, Edinburgh.

Gopalan, C., Ramasastri, B.V. and Balasubramaniam, S.C. 1999. *Nutritive Value of Indian Foods.* National Institute of Nutrition, ICMR, Hyderabad.

Gordon, M. Wardlaw and Paul, M.Insel. 1990. *Perspectives in Nutrition*, Times Mirror/Mosby College Publishing, St. Louis.

Guthrie, Helen A. and Picciano Mary Francis. 1995. *Human Nutrition.* McGraw-Hill, Boston.

Guyton, Arthur C. 1991. *Textbook of Medical Physiology*, Prism Book Pvt. Ltd., Bangalore.

Jain J. L., Sunjay Jain and Nitin Jain. 2005. *Fundamentals of Biochemistry,* S. Chand and Company Ltd., New Delhi.

Lehninger, 1987. *Principles of Biochemistry.* CBS Publishers and Distributors, New Delhi.

Murray, R.K., Granner, T.K., Mayes, T.K. and Rodwell, V.W. *Harper's Biochemistry,* 21st edn. McGraw-Hill, New Delhi.

Muthayya, N.M. 1991. *Essentials of Physiology.* Vols I and II. Emerald Publishers.

Nath, R.L. 1996. *Textbook of Medicinal Biochemistry.* New Age International, New Delhi.

Passmore, R. and Eastwood, M.A. 1987. *Davidson and Passmore's Human Nutrition and Dietetics.* ELBS, Churchill, Livingstone.

Rastogi, S.C. 1993. *Biochemistry.* TataMcGraw-Hill Publishing Company Ltd., New Delhi.

Robinson, C.H. and M.R. Lawler, 1982. *Normal and Therapeutic Nutrition*, Oxford and IBH Publishing Co., New Delhi.

Satyanarayana, 2002. *Essentials of Biochemistry*. Books and Allied (P) Ltd., New Delhi.

Shils, M.E., James, A. Olson and Moshe Shike. 1999. *Modern Nutrition in Health and Disease.* Lea and Febizer, Philadelphia.

Swaminathan, M. 1988. *Essentials of Food and Nutrition*. Vol I and II. The Bangalore Printing and Publishing Co. Ltd., Bangalore.

Weil, J.H. 1990. *General Biochemistry,* 6th edn. Wiley Eastern Ltd. New Delhi.

William, Sue Rodwell. 1985. *Nutrition and Diet Therapy*, Times Mirror/Mosby College Publishing, St. Louis.

www.nutrition.org

www.fascb.org/asns/intro.html

www.eatright.org

arborcom.com

www.wadsworth.com/nutrition

INDEX